<u>Disclaimer</u>

Book Title: Simulation of the Dynamics of a Wind-Driven Fire in a Ranch-Style House - Texas

Book Author: Adam M. Barowy; Daniel Madrzykowski;

Book Abstract: This report describes the results of calculations using the NIST Fire Dynamics Simulator (FDS) performed to provide insight on the thermal conditions that may have occurred during a wind-driven fire in a one-story ranch house on April 12, 2009 in Houston, Texas. The FDS simulations represented the building geometry, material thermal properties, and fire behavior based on information gathered from multiple sources. The simulation results are provided in this report. The FDS simulation that best represents the witnessed fire conditions indicate that fire spread throughout the attic and first floor developed a wind-driven flow with temperatures in excess of 260 °C (500 °F) between the den and front door. The critical event in this fire was the creation of a wind-driven flow path between a large span of failed windows on the upwind side of the structure, and the open front door on the downwind side of the structure. Floor-to-ceiling temperatures rapidly increased in the flow path, in which members were performing interior operations. In a simulation without wind, the flow path was not created after the large span of windows failed, and the thermal environment surrounding the location of interior operations improved.

Citation: NIST TN - 1729

Keywords: cfd models; fire dynamics; fire fatalities; fire fighters; fire investigations; fire models; fire simulations; wind driven fire

NIST Technical Note 1729

Simulation of the Dynamics of a Wind-Driven Fire in a Ranch-Style House - Texas

Adam Barowy
Daniel Madrzykowski

NIST National Institute of Standards and Technology • U.S. Department of Commerce

This page intentionally left blank

NIST Technical Note 1729

Simulation of the Dynamics of a Wind-Driven Fire in a Ranch-Style House - Texas

Adam Barowy
Daniel Madrzykowski
Engineering Laboratory
Fire Research Division
National Institute of Standards and Technology

January 2012

U.S. Department of Commerce
John E. Bryson, Secretary

National Institute of Standards and Technology
Patrick D. Gallagher, Under Secretary of Commerce for
Standards and Technology and Director

<u>Use of Non-SI Units in a NIST Publication</u>

It is NIST policy to use the International System of Units (metric units) in all its publications. In this report, however, information is presented in U.S. Customary Units (inch-pound), as this is the preferred system of units in the U.S. construction industry and fire service.

Certain commercial software, equipment, instruments, or materials may have been used in the preparation of information contributing to this report. Identification in this report is not intended to imply recommendation or endorsement by NIST, nor is it intended to imply that such software, equipment, instruments, or materials are necessarily the best available for the purpose.

National Institute of Standards and Technology Technical Note 1729
Natl. Inst. Stand. Technol. Tech. Note 1729, 86 Pages (January 2012)
CODEN: NTNOEF

Abstract

This report describes the results of calculations that were performed using the Fire Dynamics Simulator (FDS) to examine the effect of wind on the thermal and fire conditions in a single-story residential structure and to provide a visualization of fire behavior representative of what members of the Houston Fire Department (HFD) experienced during the course of their interior operations, to provide insight on the thermal conditions that claimed the lives of two fire fighters during a wind-driven fire in a one-story ranch house on April 12, 2009, in Houston, Texas. FDS simulations were developed that represented the building geometry, material thermal properties, and fire behavior based on information and photographs from the investigation report from the Texas State Fire Marshal's Office, information from the National Institute for Occupational Safety and Health (NIOSH) fire fighter fatality investigation report, a detailed floor plan provided by the Bureau of Alcohol, Tobacco, and Firearms (ATF), and additional photos, video and dimensional measurements collected during a site visit by National Institute of Standards and Technology (NIST) staff. The results from the simulations are provided in this report.

The FDS simulation that best represents the witnessed fire conditions indicates that the fire that spread throughout the attic and first floor developed a wind driven flow with temperatures in excess of 260 °C (500 °F) between the den and front door. The critical event in this fire was the creation of a wind-driven flow path between the upwind side of the structure and the exit point on the downwind side of the structure, the front door. The flow path was created by the failure of a large span of windows in the den, in the rear of the structure. Floor-to-ceiling temperatures rapidly increased in the flow path where multiple crews were performing interior operations. In a simulation that excluded wind, the flow path was not created, and the thermal environment surrounding the location of interior operations was improved.

Wind has been recognized as a contributing factor to fire spread in wildland fires and large-area conflagrations and wildland fire fighters are trained to account for the wind in their tactics. While structural fire departments have recognized the impact of wind on fires, in general, the standard operating guidelines for structural fire fighting have not changed to address the hazards created by a wind driven fire inside a structure. The results of the "no-wind" and "wind" fire simulations demonstrate how wind conditions can rapidly change the thermal environment from tenable to untenable for fire fighters working in a single-story residential structure fire. The simulation results emphasize the importance of including wind conditions in the scene size-up before beginning and while performing fire fighting operations and adjusting tactics based on the wind conditions. These results are in agreement with NIST studies conducted to examine wind driven fire conditions in high-rise structures.

Key Words: cfd models; fire dynamics; fire fatalities; fire fighters; fire investigations; fire models; fire simulations; wind driven fire

This page intentionally left blank

Table of Contents

List of Figures

List of Tables

This page intentionally left blank

1. Introduction

Part of the mission of the Fire Research Division at the National Institute of Standards and Technology (NIST) is to develop and apply technology, measurements and standards, improve the understanding of the behavior, prevention and control of fires to enhance fire fighting operations and equipment, fire suppression, fire investigations, and disaster response. NIST has previously used the Fire Dynamics Simulator (FDS) to provide insight on the fire development and thermal conditions of other multiple-fatality fires [1,2,3,4,5,6,7]. The overriding objective of all of these studies is to improve the safety of fire fighters.

On April 12, 2009, a fire in a one-story ranch home in Texas claimed the lives of two fire fighters. Sustained high winds occurred during the incident. The winds caused a rapid change in the dynamics of the fire after the failure of a large section of glass in the rear of the house. At the request of NIOSH and the Houston Fire Department (HFD), NIST assisted with examining the fire dynamics of this incident. NIST performed computer simulations of the fire using the FDS and Smokeview, a visualization tool, to provide insight on the fire development and thermal conditions that may have existed in the residence during the fire.

The specific objectives of the simulations detailed in this report are:

1) To examine with physics-based calculations the effect of wind on the thermal and fire conditions in this single-story residential structure.
2) To provide visualization of fire behavior representative of what HFD members experienced during the course of their interior operations.

This document describes the input and the results of two FDS (version 5.5.3) simulations. One simulation is intended to demonstrate wind-driven fire conditions in the structure, and the second simulation is intended to show how the fire may have occurred in the absence of wind.

Wind has been long recognized as a contributing factor to fire spread in wildland fires and large-area conflagrations. The Fire Department of New York City (FDNY) identified that wind conditions significantly increased the severity of high-rise fires and that wind-driven fires were challenging their resources, tactics, and safety. As part of a collaborative effort with FDNY, the Chicago Fire Department, the Fire Protection Research Foundation, Polytechnic Institute of New York University, and fire chiefs from fourteen North American fire departments, NIST completed fire experiments in the laboratory and in a 7-story structure to enable a better understanding of wind-driven fire tactics, including structural ventilation and suppression [8,9]. As part of the laboratory experiments it was found that wind-driven fire behavior can occur with wind speeds as little as 4.5 m/s (10 mph). Wind-driven fire behavior also has been recognized in other types of structures, including single-family homes [10].

While structural fire departments have recognized the impact of wind on fires, in general, the standard operating guidelines for structural fire fighting have not changed to address the hazards created by a wind driven fire inside a structure. The results of the "no-wind" and "wind" fire simulations demonstrate how wind conditions can rapidly change the thermal environment from tenable to untenable for fire fighters working in a single-story residential structure fire. The simulation results emphasize the importance of including wind conditions in the scene size-up before beginning and while performing fire fighting operations and adjusting tactics based on the wind conditions. These results are in agreement with NIST studies conducted to examine wind driven fire conditions in high-rise structures.

2. Fire Incident Summary

The following account of events is based on information provided in the Texas State Marshal's report [11] and the National Institute for Occupational Safety and Health (NIOSH) report [12]. Details regarding distances and times were taken from these two reports and should be regarded as approximations. A summary of the events describing the conditions of the fire and fire department operations is provided in Table 1: was developed using the information from the NIOSH prepared timeline.

At approximately midnight on April 11, 2009, a fire was observed in a closet near the master bedroom in a single story, ranch-style home. Figure 1 shows an aerial view of the structure, including the direction of the wind, the direction of apparatus travel, and numerous features of the property that hindered efforts to complete a 360° survey of the structure. Figure 1 also shows the direction and velocity of the high winds, which carried optically dense smoke exhausting from the structure into the path of apparatus travel.

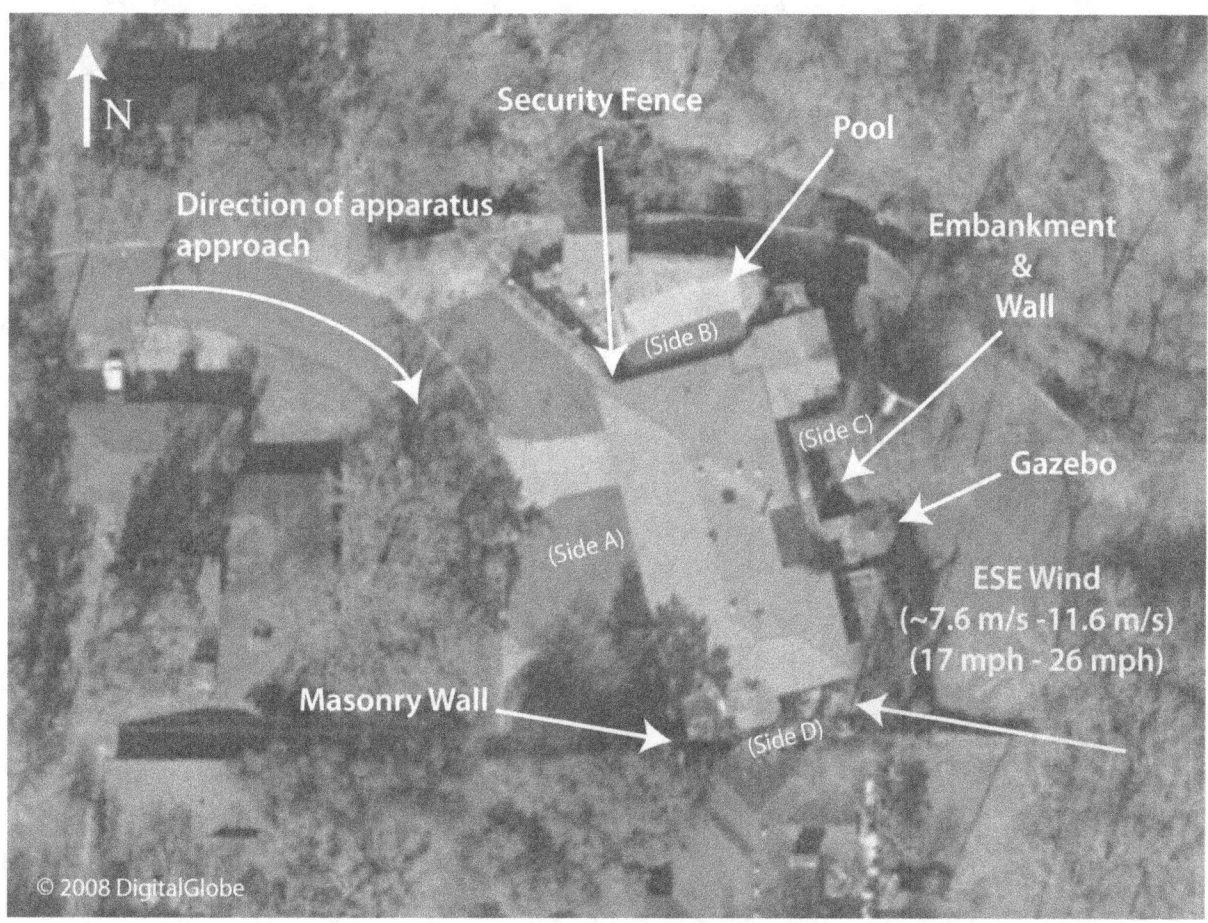

Figure 1: Aerial view of residence. Background image used with permission of Digital Globe. Enhancements by NIST.

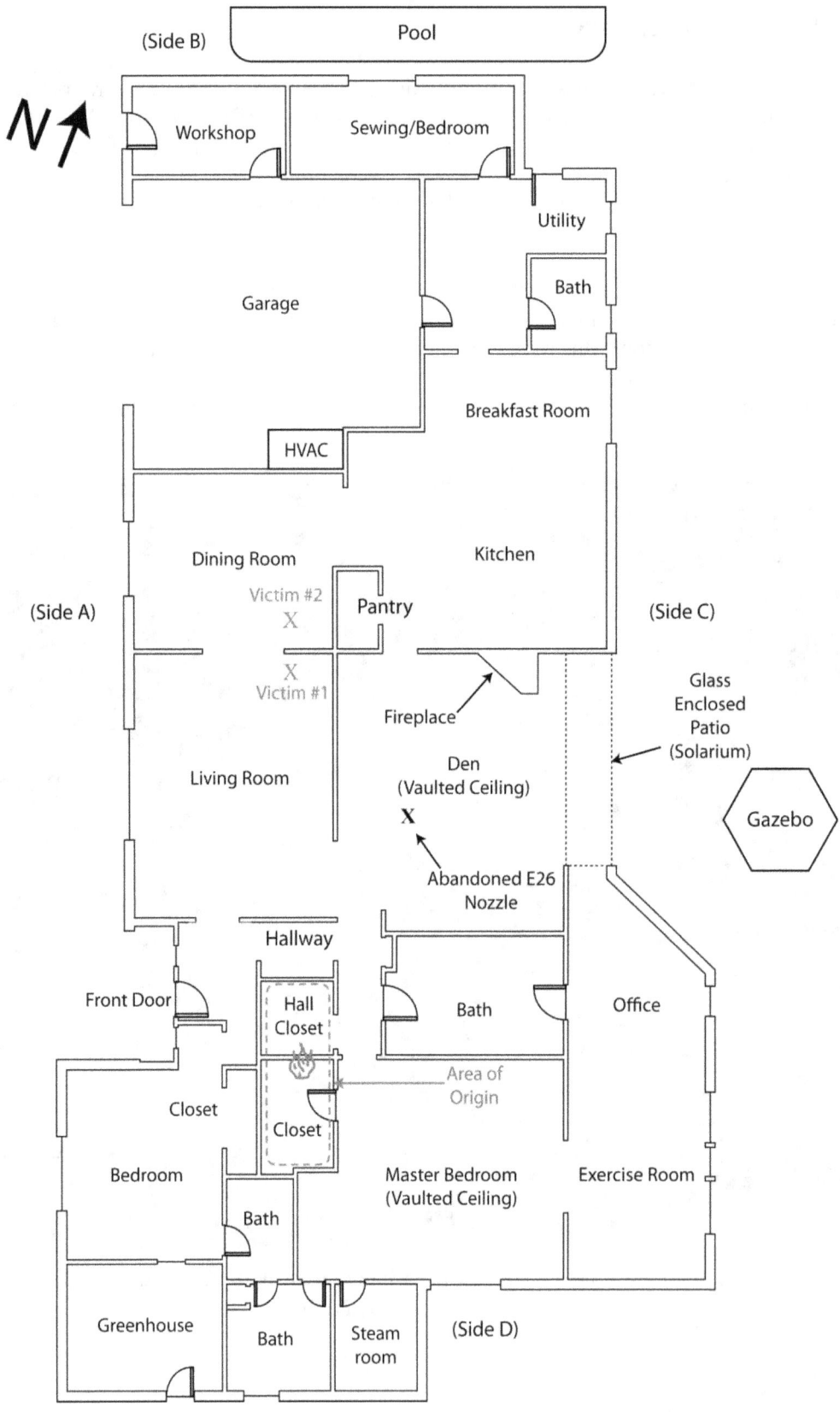

Figure 2: Plan view of residence. Drawn using measurements collected by NIST staff. Locations of victim #1 and victim #2 and nozzle from NIOSH report [12].

According to the Texas State Fire Marshal's investigation, the fire was ignited by overheated wiring in the ceiling light fixture. Figure 2 shows a plan view of the structure.

It is unknown how long the fire had been burning before being detected by the occupants, who were awake at the time. While preparing for bed, an occupant in the master bedroom observed a flickering light coming from the closet. Upon inspection, flames and smoke were visible from "eye level to ceiling level inside the closet" [11]. After being alerted, the second occupant walked from the kitchen toward the hallway and observed smoke and flames exiting the closet doorway. As the occupants exited, the interior and exterior overhead garage doors were left open. Eventually, the occupants saw flames exiting the roof of the structure, near the skylight in the bathroom across from the closet where the fire originated. Figure 2 shows a plan view of the residence with the location of the area of fire origin marked.

Just before 12:08 AM on April 12, HFD dispatch received a report from a neighbor reporting that a house was on fire and that there was "smoke all over the place."
By 12:09 AM, HFD dispatch marked a first alarm assignment en route. Sustained winds blew from the ESE at approximately 7.6 m/s (17 mph), and gusted to approximately 11.6 m/s (26 mph)[1]. This wind carried optically dense smoke from the structure into the street, opposite the direction of apparatus arrival.

The first engine on-scene, Engine 26 (E-26), arrived within five minutes at approximately 12:13 A.M. As E-26 approached the scene, the smoke exhausting from the roof was of great enough quantity that the E-26 crew experienced significant difficulty locating the structure. E-26 advised dispatch that they would perform a "fast attack" (aggressive interior attack). The E-26 crew, comprised of a probationary fire fighter (victim #1), a Captain (victim #2), and third fire fighter, advanced a pre-connected hose line to the front door and forced the door open. Ladder 26 (L-26) arrived within seconds of E-26. The L-26 captain observed fire coming from a turbine vent near the peak of the roof and radioed that L-26 would be venting the roof.

Engine 36 (E-36) arrived on scene at approximately 12:15 A.M. and observed the E-26 crew advance into the structure while walking upright. The third E-26 fire fighter remained near the front door to pull slack for the hose and experienced heavy smoke and high heat conditions just inside the doorway. The E-26 crew advanced the hose line down the hallway toward the den. One wall on side C (the rear) of the structure was composed of large glass panels.

Approximately one minute after E-26 crew members entered the structure (12:16 A.M.), E-36 advanced a second hose line through the front door and down the hallway. The E-36 captain peered beneath the smoke layer and briefly saw the legs of the E-26 crew 1.5 m to1.8 m (5' to 6') ahead, advancing down the hallway until they disappeared into the smoke layer. As the E-36 crew advanced down the hallway, the helmet of E-26 fire fighter #3 was dislodged. E-26 fire fighter #3 exited the structure to replace the dislodged helmet. The fire fighter observed flames rolling overhead while exiting the front door. E-36 advanced approximately 3 m to 5 m (10 ft to 15 ft) down the hallway and began removing ceiling material to search for fire in the attic (pulling the ceiling). Suddenly, the temperature rapidly increased and forced the E-36 crew to the floor. The E-36 fire fighter directed the hose line toward a red glow observed through the hole, but encountered inadequate pressure and disrupted water flow. In response, the E-36 crew began to back down the hallway toward Side A of the structure and experienced further temperature increases accompanied by "whooshing or roaring" sounds. Upon reaching the doorway to the living room, the E-36 crew advanced in a few feet and operated the hose line on the fire overhead. Water from the 1-¾ inch hose line had little effect on the fire. This was also the approximate location of a ceiling breach made by L-26 from the roof.

---

[1] Wind data referenced from a weather station at the William P. Hobby airport, approximately 3 km (2 mi) away.

At the time that the E-36 crew entered the structure, L-26 was performing ventilation cuts on the roof. The L-26 captain observed that flames came through the cuts made by the roof saw. The L-26 captain also observed flames 0.3 m to 0.6 m (1'-2') wide lapping over the eaves in the rear of the structure, in the same location as the glass-enclosed patio. At approximately 12:19 A.M., L-26 completed the roof cuts, pried up the roof decking and used a pike pole to punch down through the first floor ceiling twice. Immediately, flames exited the vent hole in a "swirling vortex" that extended several feet into the air. At approximately the same time the vent operation was completed, the L-26 captain observed the fire overlapping the eaves on side C of the structure expand to approximately 6 m (20 ft) wide (the width of the glass-enclosed patio). The fire overlapping the eaves spread rapidly up the roof toward the location of the L-26 crew.

At approximately 12:20 A.M., as the L-26 crew descended from the roof and the E-36 continued to operate the hose line in the living room, Ladder 29 (L-29) entered the front door on orders to assist E-26 with a primary search of the structure. The L-29 crew immediately dropped into a crawl due to the high heat conditions. The L-29 crew noted thermal conditions that continued to worsen as they advanced along the E-26 hose line toward the den. At one point, the L-29 captain observed the legs of upright crew members advancing the hose line forward toward the den. The forward progress of L-29 was halted to clear debris blocking the hall. After clearing the debris, the L-29 captain observed that the hose line was no longer being advanced.

The L-29 crew found that the smoke layer dropped almost to floor level as they advanced further toward the den. Eventually, the L-29 crew discovered the abandoned nozzle of E-26 in the den and began operating the hose overhead. The E-36 and L-29 crews both operated their hose lines until incident command (IC) radioed to switch to a defensive mode at approximately 12:21 A.M because of worsening conditions and a large volume of fire exiting the roof vent. E-36 and L-29 exited from the front door at approximately 12:22, following the hose line.

At 12:22 A.M., the IC requested a personal accountability report (PAR) from all crews. The PAR indicated that all fire fighters were accounted for except for the E-26 captain and probationary fire fighter. Over the next 24 minutes, defensive operations were carried out using a ladder-pipe and 2-½ inch hose line. During this period, a rapid intervention crew (RIC) and the L-26 crew made multiple attempts to re-enter the structure but could not because of the extreme heat. At approximately 12:46 A.M., enough of the fire was extinguished so that HFD members could re-enter the structure. At 12:51 A.M., the E-26 probationary fire fighter was located in the living room. At 12:52 A.M., the E-26 captain was located in the dining room. Both fire fighters died from injuries caused by the fire.

Figure 3 and Figure 4 show the locations where the captain and probationary fire fighter were located. These pictures show that the two fire fighters moved from an area of high thermal energy (based on the location of their nozzle and statements from other fire fighters) to an area with less thermal energy (based on the amount of damage where they were found). The carpet in the areas in which the victims were found was unburned, and upholstery on the seats of the chairs was still intact; whereas the rooms outside the living room and den showed much more thermal damage. Normally, fire fighters would follow their fire hose (line) out the door to safety. In this incident, the line was in the hot gas flow path. This may be why the victims left the line behind. Additionally, the wall that separated the living room from the hallway formed two parallel paths to the front door and may have led to confusion. Figure 5 shows the charred wood framing of wall that separated the hallway from the living room as viewed from the front door.

The results of the model demonstrated the impact that wind conditions can have on the thermal environment of a single-story residential structure fire. The model simulated the wind driven fire behavior that has been measured in previous experiments in high rise structures [8,9]. The model

6

simulates how rapidly the thermal conditions became untenable, even for a fire fighter in full PPE, due to the development of a flow path of hot gasses through the structure and the introduction of wind. The results emphasize the importance in a scene size-up before beginning and while performing fire fighting operations and adjusting tactics based on the wind conditions.

Figure 3: Location of the E-26 captain in the dining room doorway

Figure 4: Location of the E-26 probationary fire fighter in the north of the living room

Figure 5: View looking directly at the short wall that separated the living room and hallway

Table 1: Abridged NIOSH approximate incident timeline

Time (min)	Fire Behavior & Fireground Operations	Incident Time (min)
12:08 A.M.	HFD dispatches call Caller: "smoke all over the place" & occupants outside house Flames visible through roof vent	0
12:13 A.M.	E-26 on scene, heavy smoke reported	5
12:14 A.M.	E-26 crew forces open front door, L-26 captain reports "heavy smoke"	6
12:15 A.M.	E-26 crew advances hose line into structure while walking upright	7
12:16 A.M.	E-36 advances hose line into the structure toward den. E-36 observes E-26 upright ~1.8 m (6') away toward Side C.	8
12:18 A.M.	Flames exit through roof saw cuts as L-26 crew operates L-26 observes flames approximately 3' wide lapping from Side C eaves.	10
12:19 A.M.	L-26 completes roof vent, breaches 1st floor ceiling. Roof conditions unsafe, L-26 exits roof E-36 punches hole in ceiling near end of hall to search for fire in attic	11
12:20 A.M.	Fire lapping from Side C eaves expands to ~20 ft wide Flames extend several feet from vent opening Temperature increases rapidly, E-36 forced to floor L-29 observes legs of crew standing toward side C in den	12
12:20 A.M.	L-29 finds and operates abandoned E-26 hose line, overhead and side to side	12
12:21 A.M.	Temperature in den increases further E-36 and L-29 operating from hallway	13
12:22 A.M.	Fireground operations switched to defensive	14
12:46 A.M.	Fire knocked down	38

3. Fire Dynamics Simulator and Smokeview

FDS is a computational fluid dynamics (CFD) model that solves a form of the Navier-Stokes equations appropriate for low-speed, thermally driven flow with an emphasis on smoke and heat transport from fires. Within a CFD model, the room or building of interest is divided into small three-dimensional rectangular control volumes or computational cells. The cells are contained together within one larger volume known as a computational domain. The CFD model computes the density, velocity, temperature, pressure and species concentration of the gas in each cell. Based on the laws of conservation of mass, momentum, species, and energy, the model tracks the generation and movement of fire gasses. FDS utilizes material properties of the furnishings, walls, floors, and ceilings to compute fire growth and spread. One of the most important advantages of FDS is that it is verified and validated against fire test data to ensure that it provides the expected results, given sufficient input data. A complete description of the FDS model is given in reference [13].

Smokeview is a software tool designed to visualize the results of the FDS. Smokeview visualizes smoke and other attributes of the modeled fire using traditional scientific methods such as displaying tracer particle flow, two dimensional (2D) or three dimensional (3D) shaded contours of gas flow data such as temperature and flow vectors showing flow direction and magnitude. Smokeview also visualizes fire attributes realistically so that one can visually experience the fire. This is done by displaying a series of partially transparent planes where the transparencies in each plane (at each grid node) are determined from soot densities computed by FDS. Smokeview also visualizes static data at particular times using 2D or 3D contours of data such as temperature and flow vectors showing flow direction and magnitude.

4. Model Input Parameters

FDS requires the following inputs: the geometry of the building compartments being modeled, the computational cell size, the location of the ignition source, the heat release rate (HRR) of the ignition source, physical and thermal properties of walls and furnishings, and the size, location, and timing of vent openings to the outside, which critically influence fire growth and spread.

Two simulation cases will be presented. The first simulation was completed to demonstrate fire dynamics in the structure in the absence of a wind condition at the time of the incident. The second simulation is used to contrast how the fire dynamics was affected with the addition of wind, and to provide insight into the fire environment experienced by HFD members. The additional simulation setup necessary to incorporate wind is detailed in Section 6.6.

Forensic reconstructions require the model to simulate an actual fire based on inputs developed by the user from information that is collected after the event (e.g. eyewitness accounts, burned and unburned materials, physical dimensions, etc.). The purpose of the simulation is to connect a sequence of discrete observations with a continuous description of the fire dynamics. In general, forensic reconstructions can be more challenging to perform than design or experimental applications because there is a great amount of uncertainty in the total HRR as the fire spreads from object to object. Uncertainties in material properties and measurements, fire growth, and simplifying assumptions in the model, often force the comparison between model and evidence to be qualitative.

A reconstruction is an example of an "ill-posed' problem. The outcome is known in advance whereas the pre-fire and boundary conditions, which are discussed in the following sections, are often not well known. Subsequently, there is no single unique solution to the problem, and it is possible to simulate numerous fires that produce the given outcome. There is no right or wrong answer. Instead, there are a small set of plausible fire scenarios that are consistent with the collected evidence. These simulations are then used to demonstrate why the fire behaved as it did based on the current understanding of fire physics incorporated in the model [5].

It should be considered that the application of water would have had some impact on the simulated fire and smoke conditions. However, hose stream suppression was not included in the simulations, given the fact that two 1 ¾" hose lines' operation did not significantly reduce the severity of the fire.

4.1   Geometry of Structure

The floor plan of the one-story ranch-style house involved in this incident is shown in Figure 2. Based upon the NIOSH and Texas State Fire Marshal reports, the original structure was built in 1956 [11,12]. The structure measured approximately 33 m (108 ft) long by 14.6 m (48 ft) wide, making up approximately 390 m^2 (4, 200 ft^2) of living space. The structure consisted of traditional stick-built wood construction with a brick veneer over the exterior walls. The house rested on a concrete slab foundation. The interior finish of the structure was gypsum board walls and ceilings.

Permit records indicate multiple modifications to the property, including 29.8 m^2 (320 ft^2) and 30.1 m^2 (324 ft^2) additions.   Portions of the roof were constructed over the original roof, which added to the combustible load in the attic, as well as created void spaces for pyrolized fuel to collect. The roof was constructed with asphalt shingles over wooden planks that, based on NIST measurements, were approximately 152.4 mm (6 in) wide by 19 mm (0.75 in) thick. The rafters were constructed from dimensional 2 x 6 lumber, spaced 0.61 m (24 in) on center. Figure 6 shows a charred rafter located above the dining room, with top-down charring that indicates the intensity of the fire that burned in the attic.

Figure 6: Charred rafter located above the dining room

Other known modifications that contributed to the fire dynamics during the incident included the addition of vaulted ceilings over the den and master bedroom, a skylight in the ceiling of the hallway bathroom, a bathroom on the south side, and a wall of glass panels that replaced the original east exterior wall in the den. The specific details of all the modifications are not known due to the extensive damage of the structure. Figure 7 shows an aerial view of the fire damage to the structure.

Gazebo

Detached Guest Apartment

© 2009 Houston Fire Department

Figure 7: Aerial view of damage to the structure. Image used with permission of the Houston Fire Department [11]. Enhancements by NIST. Note: Guest apartment is detached and not shown in Figure 2.

12

The physical dimensions of the structure were defined within a computational domain sized 38.7 m (127 ft) long by 21.0 m (68.9 ft) wide by 6.3 m (20.7 ft) high. The boundaries of the domain were spaced a minimum of a meter away from all exterior sides of the structure to ensure adequately resolved buoyancy driven fluid flow into and out of the structure. Figure 8 shows the structure within the computational domain.

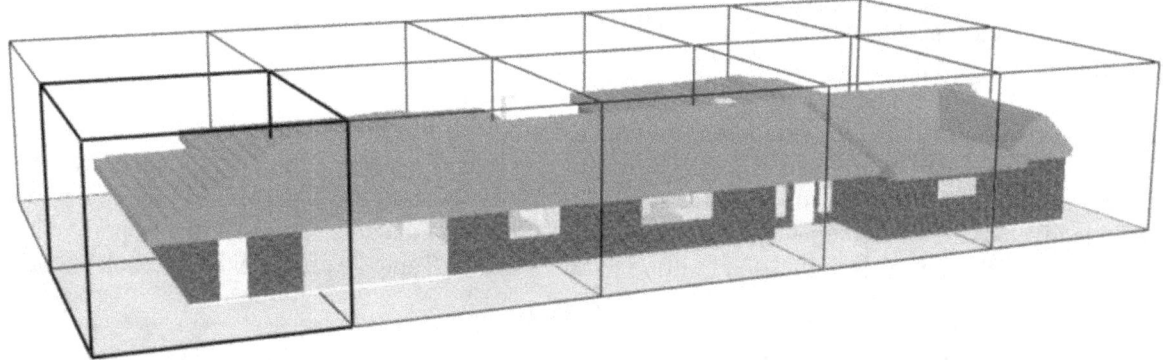

Figure 8: Computational domain of the simulation

The computational domain was divided into ten, equally sized computational meshes to take advantage of the parallel processing capability of FDS. Each "box" in Figure 8 represents an individual mesh. The bottom boundary of the computational domain was specified to be nonreactive in the simulation. The remaining boundaries were specified as "open." Open boundaries allow heat and smoke to exit the domain and make-up air to flow into the domain, which best simulates the area beyond the computational domain in this case. The specified computational cell size was 100 mm (4 in). One hundred millimeters was chosen as the computational cell size based on favorable results obtained in the simulation of the Station Nightclub Fire, which was a similarly-sized structure [6]. Figure 9 shows the grid resolution laid over a portion of the structure for perspective. The computational domain totaled roughly five million computational cells.

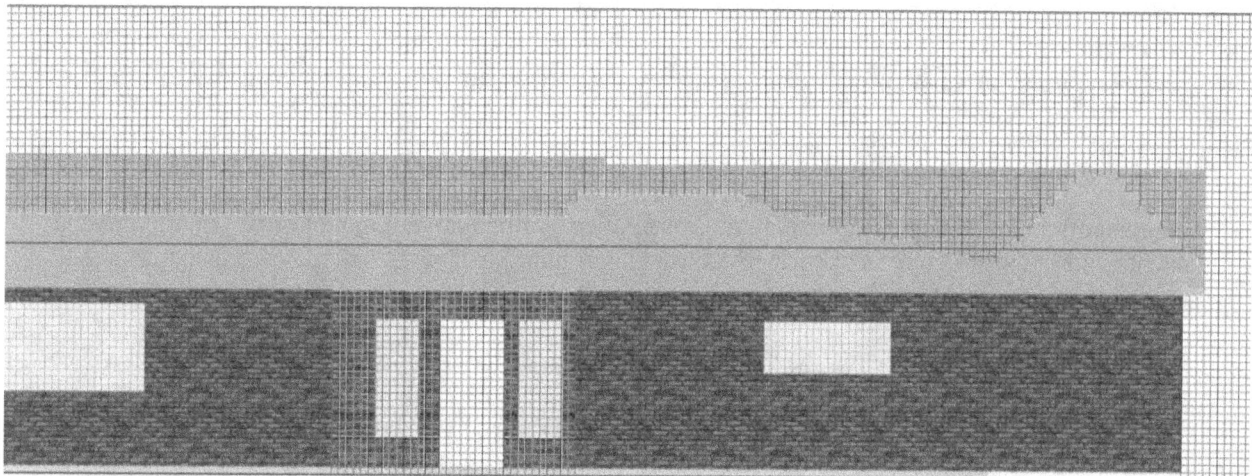

Figure 9: Partial isometric view of the front of the structure with grid cell size overlaid

4.2 Materials

When a wall, ceiling, truss, or any other structural item or piece of furniture is defined in a FDS simulation, the user prescribes physical and thermal properties that the model uses to calculate its interaction with the fire environment. Some of these properties, such as thermal diffusivity, thermal conductivity, density and thickness, influence the heat transfer in the material. For materials that burn, additional parameters such as ignition temperature, heat of combustion, heat of vaporization and maximum burning rate are specified.

A list of the materials used in the simulations conducted in this study, along with the physical and thermal properties required by FDS, is given in Table 2. Multiple references were required to gather all of the necessary thermal and physical properties for some materials (e.g., cotton). The primary source for each material is given in the first column. References are given next to the value for a specific property where necessary. Only the materials with listed ignition temperatures in Table 2 were considered fuels in the model.

Table 2: Thermo-physical properties of materials used in simulation

Material	Ignition Temp. (°C)	Peak Heat Release Rate Per Unit Area[2] (kW/m^2)	Heat of Combustion (kJ/kg)	Thermal Conductivity at 20° C (W/(m·K))	Density (kg/m^3)	Specific Heat at 20° C (kJ/(kg·K))
Brick [14]	N/A	N/A	N/A	0.69	1600	0.84
Carpet [15]	N/A	N/A	N/A	0.16	750	4.50
Cotton	260 [16]	180 [17]	15,600 [18]	0.11 [19]	36.4[3]	1.29 [19]
Douglas Fir [20]	384	181.40	13,000 [21]	0.13	502	1.8 [22]
Fiberglass Insulation	N/A	N/A	N/A	0.04 [23]	14 [24]	0.84 [14]
Glass [14]	N/A	N/A	N/A	0.76	2700	0.84
Gypsum [25]	400 [26]	224 [27]	5,600 [21]	0.17	930	1.09

All wood components (e.g., rafters, sheathing) in the simulation were modeled as Douglas fir. Gypsum surfaces were modeled to allow the painted paper face to contribute to the fuel load. Table 3 details the arrangements and thicknesses of the materials as applied to structural components of the structure and contents within the structure. For surfaces consisting of multiple layers, materials are listed from the interior to the exterior of the living space.

The hallway closet is described in the Texas State Fire Marshal's report as having a cedar plank veneer over gypsum wallboard. The closet contained linens, bedding materials, pillows, and other combustibles [11]. A volume of 1.9 m^3 of cotton material was modeled on wooden shelves to approximate the load of combustible materials stored in the closet. This closet fuel load and associated ignition source used to ignite it were sufficient to provide fire conditions similar to those described by the occupants at the time they discovered the fire and to contribute to the buildup of heat and smoke leading to the conditions described by HFD.

---

[2] Heat release rate per unit area was input as time varying function. The cone calorimetry data used in conjunction with the peak HRR is provided in Appendix A: FDS Input File.
[3] Cotton density was estimated with cotton towel data from [40] and an estimated thickness of 2.5 mm for a compressed towel.

Table 3: Materials as applied to modeled components

Material Application	Type of Material(s)	Thickness
Carpeted Floor	Carpet	0.006 m (0.24 in)
Ceilings	Gypsum, Fiberglass Insulation	0.013 m (0.50 in), 0.090 m (3.50 in)
Exterior Walls	Gypsum, Brick	0.013 m (0.50 in), 0.090 m (3.50 in)
Fireplace	Brick	0.090 m (3.50 in)
Garage Floor	Inert	N/A
Hall Closet Contents	Cotton	0.10 m (3.94 in)
Interior Walls	Gypsum	0.013 m (0.50 in)
Roof Sheathing	Douglas Fir	0.019 m (0.75 in)
Roof Trusses	Douglas Fir	0.038 m (1.50 in)
Tile Solarium Floor	Inert	N/A
Other Furniture	Inert	0.10 m (3.94 in)
Walls in Hall Closet	Douglas Fir, Gypsum	0.003 m (0.12 in), 0.013 m (0.50 in)
Windows	Glass	0.005 m (0.20 in)
Wood Floor	Douglas Fir	0.019 m (0.75 in)
Wood Furniture	Douglas Fir	0.038 m (1.50 in)

It was determined that the den contained upholstered furniture as well as additional common household items that contribute to fuel loading (e.g., area rugs, lamps, end tables, etc.). The quantity of fuel in the den was burning significantly enough that the E-26 crew was operating their hose line in the den. The contents of the den were approximated with two upholstered couches on a wood floor. The couches were modeled with gas burners applied to the largest faces, as shown in Figure 10. The burner surfaces were modeled using HRR data collected from a couch burned in a single room enclosure in the NIST Large Fire Laboratory. The HRR data used as input is provided in Appendix A: FDS Input File. Although this estimate may be much less than the actual combustible load that existed at the time of the incident, it provides adequate fuel to simulate the fire conditions described in the NIOSH and Texas Fire Marshal's reports [11,12]. Other items of furniture visible in the simulation were modeled as nonreactive obstructions for visual purposes and do not contribute to the model results. The carpeted flooring found in the structure did not burn, so it was not included in the simulated combustible load.

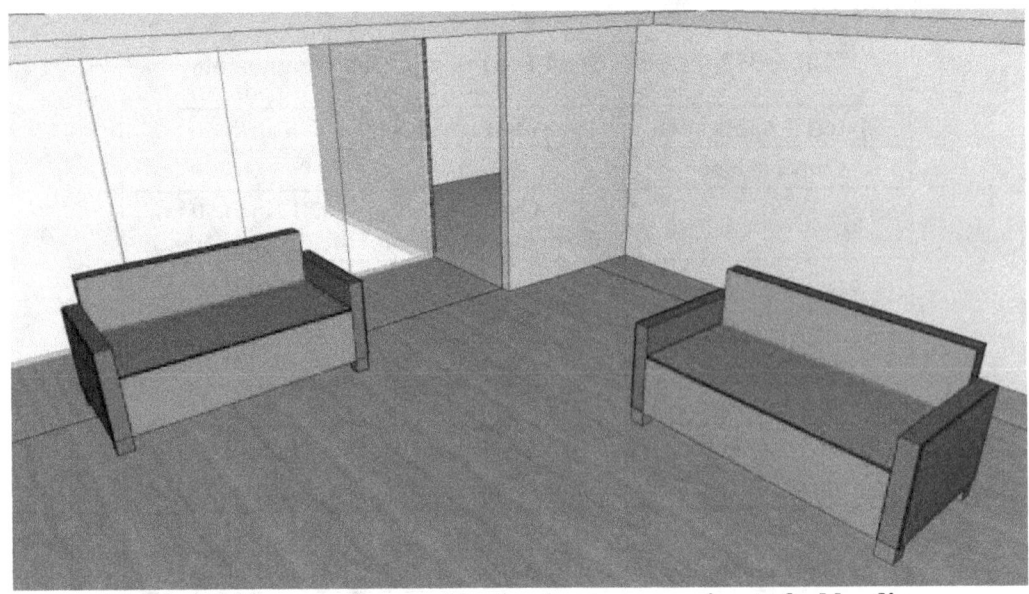
Figure 10: Two couches used in the den to approximate fuel loading.

4.3 Vents

For the purpose of this simulation, a vent is considered an opening that will allow fire gasses and ambient air to communicate between spaces within the structure as well as between spaces in the structure and the environment outside the structure. This simulation included vents to account for ventilation designed as part of the structure (e.g., heating, ventilation and air conditioning vents (HVAC), attic vents), the approximate leakage of the building envelope and changes in ventilation due to a combination of fire department operations and fire acting on the structure. The times of vent openings due to the latter are given in Table 4.

HVAC supply registers were modeled by creating penetrations from the living space into the attic. The registers were included in the model to allow communication of heat and smoke between the living space and the attic. A single hole, sized 0.2 m (7.8 in.) by 0.1 m (3.9 in.), was used in multiple locations to represent HVAC supply registers. The location and size of supply registers was approximated by reconstructing the HVAC system from scene photographs.

The total amount of attic ventilation was estimated by the using the ventilation requirements of the International Residential Code[4] (IRC) [28]. The IRC requires that the total net free ventilating area is not less than 1/150 of the area of space ventilated and that at least 50 % of the ventilation is dedicated to exhaust. The estimated area of the space ventilated in the attic is 571 m^2 (6144 ft^2), which results in a calculated 3.8 m^2 (40.1 ft^2) of net free ventilating area for the attic. It was assumed that turbine ventilators mounted on the roof provided 50 % of the attic ventilation. Figure 11 is an aerial view of the roof that shows five turbine ventilators, a roof access, the chimney and two unidentifiable smaller style vents.

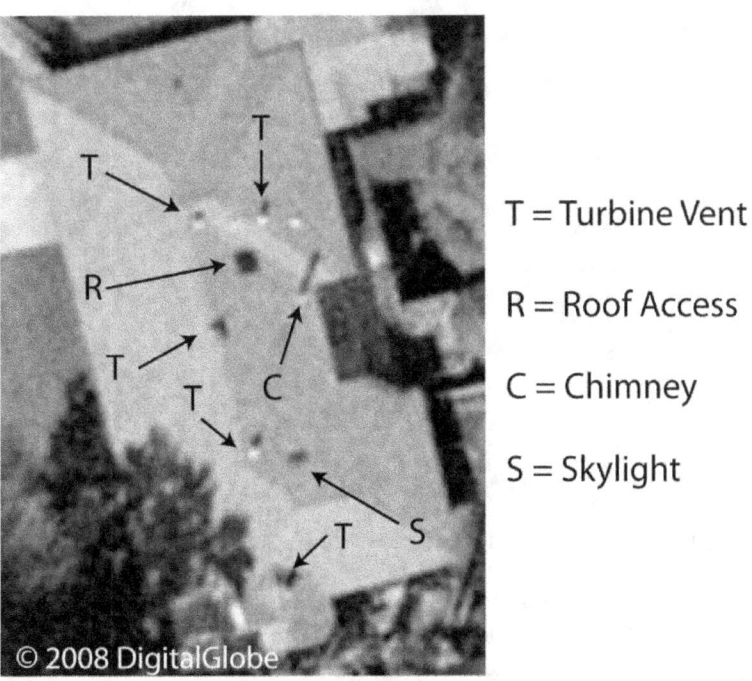

T = Turbine Vent

R = Roof Access

C = Chimney

S = Skylight

Figure 11: Aerial view of the roof identifying ventilators, skylight, roof access and chimney. Background image used with permission of Digital Globe. Enhancements by NIST.

---

[4] While the 2006 version of the International Residential Code is cited, this requirement was promulgated by the National Housing Authority in 1942 [42]. The structure concerned in this report was constructed in 1956.

The turbine ventilators were modeled as 0.4 m (15.7 in) by 0.4 m (15.7 in) open holes to represent typical 0.36 m (14 in) sized ventilators in a no-wind condition. One size of soffit vent, measuring 2.1 m (6.9 ft) by 0.1 m (0.3 ft) was assumed representative of all the soffit vents in the structure; much of the overhanging roof was burned away. The soffit vent size was approximated based on an intact vent on side B of the structure, as shown in Figure 12. Nine soffit vents were defined in the model to provide a total intake free ventilating area of 1.9 m^2 (20.5 ft^2).

Figure 12: Soffit vent on the B-side (North) of the structure

Typically, structures such as this have some amount of leakage through cracks and seams in the building envelope. Since physical objects can only be modeled with the resolution of the grid, small leaks may be created by either using a very small grid size, or by representing many small leaks by fewer large leaks. Using a smaller grid size for this simulation caused impractically long calculation times. Leakage from the structure was modeled by removing 0.1 m (0.3 ft) across the width of the bottom of the front and North (side B) doors, resulting in a leakage area of 0.18 m^2 (1.94 ft^2). This leakage area is similar to the amount found in pressure tests of houses from the same period [29].

During the course of the incident, multiple vents were opened. The times of vent openings are provided in Table 4. The timing and locations of three of the vent openings, the front door, the roof vent, and the hole created by E-36, are based on HFD radio traffic as reported in [11] and [12].

Table 4: Summary of vent times

Incident Time (min)	Fire Behavior	Simulation Time[5] (min)	Simulation
-	Flaming combustion in attic and closet Occupants exit, leaving garage door open	0	Ignition sources activated in attic & closet
0	Fire incident dispatched	2.3	-
6	E-26 forces open front door	8.3	Front door opened
7.7	Partial failure of vaulted ceiling in den	10	Vent opened in den ceiling
9.7	L-26 observes flames approximately 1 m (3 ft) wide lapping over side C eaves.	12	One horizontally oriented pane of solarium glass removed
11	L-26 completes roof vent, breaches 1st floor ceiling. E-36 pulls ceiling near end of hall	13.3	Roof vent opened/Ceiling holes added underneath vent Hole added in ceiling for E-36 vent
12	Fire lapping from Side C eaves expands to approximately 6 m (20 ft) wide	14.3	Remaining solarium glass removed
14	IC orders operations to switch to defensive	16.3	**End of Simulation**

The times of the partial failure of the den ceiling, failure of one solarium window pane, and total failure of the solarium glass are estimated based on the fire behavior observations in the NIOSH report [12] and from HFD members [30]. All of the vents are shown in Figure 13.

---

[5] Direct comparison of simulation conditions with the actual incident conditions begins after 140 seconds of simulation time. This time was used to simulate the heat and smoke filling of the attic and first floor that occurred before crews were dispatched to the incident.

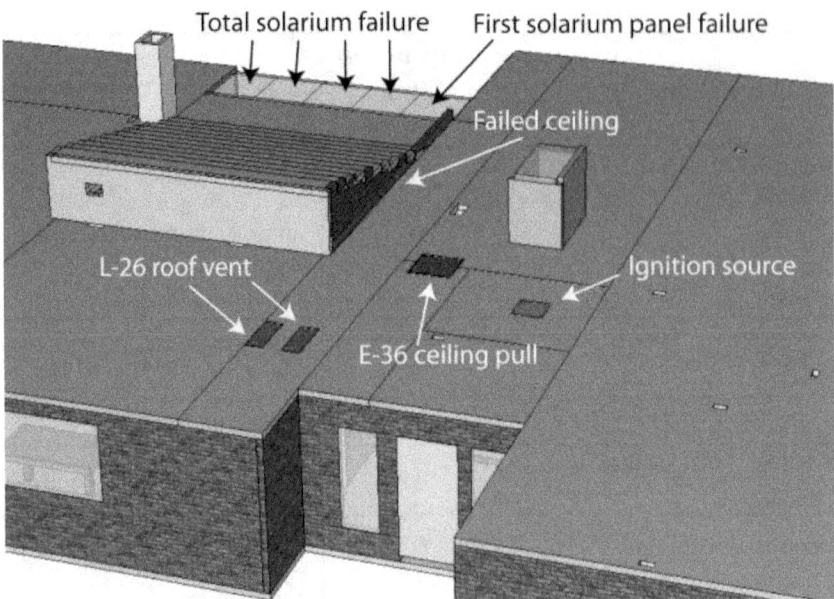

Figure 13: View of the attic, with the roof and rafters cut away. Simulated vent locations are labeled. The ignition source is shown in red.

A vent between the attic and the den is opened after 10 simulation minutes to simulate a partial failure of the gypsum wall board adjacent to the vaulted ceiling in the den. Above 100 °C, absorbed and chemically bound water is driven out of gypsum. The loss of moisture causes shrinkage and cracking. Above 300 °C, the paper facing that contains the gypsum core burns off, which is the primary factor controlling the failure of gypsum board [26]. In previous experiments, gypsum ceilings started to collapse after 7 min of fire exposure on one side, with up to 50 % of the ceiling collapsing in a particular room after 8 min of exposure [31]. This area of the vaulted ceiling was subject to thermal exposure on both sides. Fire spread through the attic, and the plume of fire gasses exited the closet and flowed over this area into the den. Figure 14 shows the den ceiling vent in blue and the position of the closet down the adjacent hallway.

The failure of the solarium glass windows were timed based on the observations of the L-26 captain, who observed flames approximately 0.3 m to 1.0 m (1 ft to 3 ft) wide overlapping the eaves at the rear (East) of the structure approximately 10 minutes after the incident dispatch. This was simulated by removing the first panel of glass marked in Figure 14. About two minutes later, the L-26 captain observed the flaming area to expand to approximately 6 m (20 ft) wide. Subsequently the remainders of the glass panels were removed after 14.3 simulation minutes.

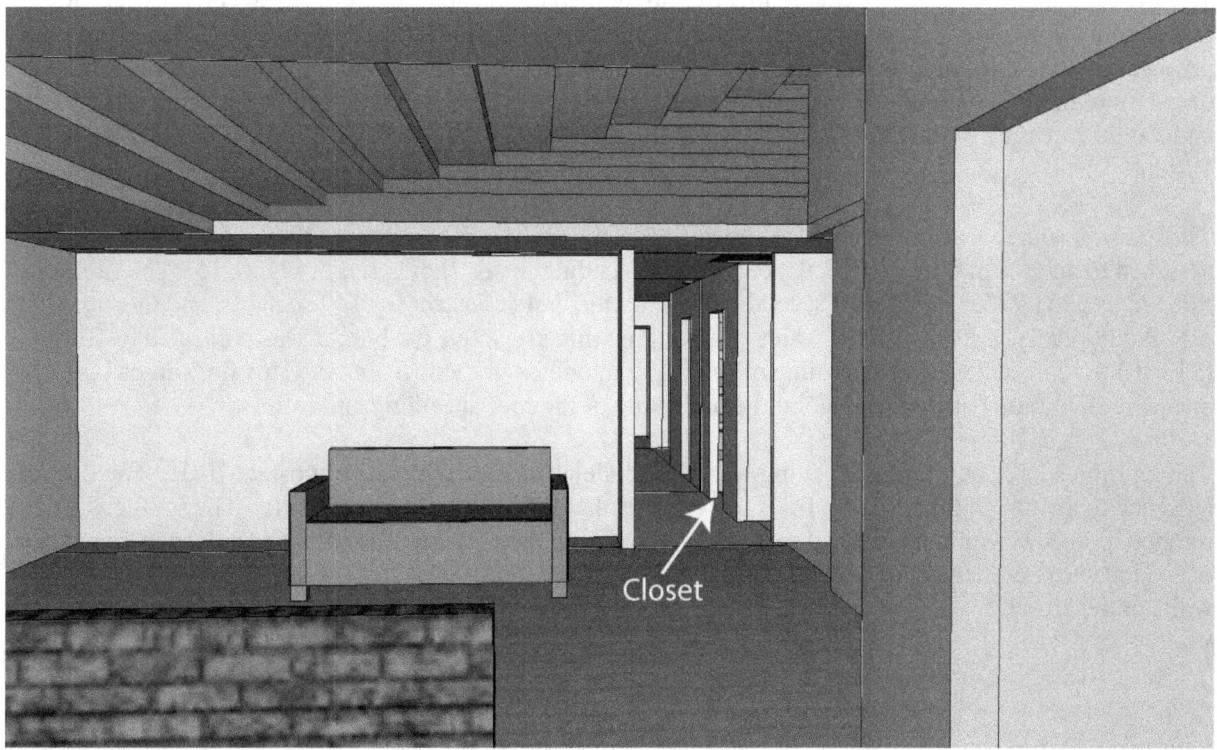

Figure 14: View of the simulated gypsum failure area (blue) from behind the brick knee wall in the kitchen.

4.4   Ignition Sources

According to the Texas State Fire Marshal's report [11], the source of fire ignition was a "loose connection" in the porcelain light fixture in the bedroom. The loose connection "resulted in a glowing connection, causing localized heating that ignited adjacent combustible materials" [11]. The actual fire may have taken several hours to develop to the flaming stage given the mechanism of ignition.

The FDS model is not used to explain how the fire ignited or spread to the attic. The complex physics and chemistry involved with overheated wiring connections that transition to flaming ignition occur on a physical scale that is much smaller than the computational cell size used in the FDS simulations. Instead, the fire simulation was initiated with two flaming ignition sources, on either side of the ceiling, to simulate the heating of the light fixture that eventually transitioned into flaming fires in the attic and closet.

The ignition source on the attic side was specified as a burner with a constant HRR. The fire was assigned to an area of 0.25 m^2 (2.7 ft^2) between two ceiling joists, shown red in Figure 15. The fire was ramped to 250 kW for the first 10 seconds of simulation, held constant for 140 seconds, and then turned off. Additionally, a 2.1m^2 (22.6 ft^2) area of sheathing directly above the burner was preheated to 100 °C (212 °F) to represent heating over time of the material bathed in the fire plume. This fire source was adequate to initiate flame spread along the underside of the roof sheathing and rafters.

The ignition source on the closet ceiling was also modeled as a burner with a constant HRR. The fire was assigned to an area of 0.25 m^2 (2.7 ft^2) directly beneath the fire initiating in the attic. The source was ramped to 70 kW within the first 10 seconds of simulation, held constant for 100 seconds, and then turned off. This fire was adequate to initiate burning on the top of the storage shelves and the wood veneered walls in the closet.

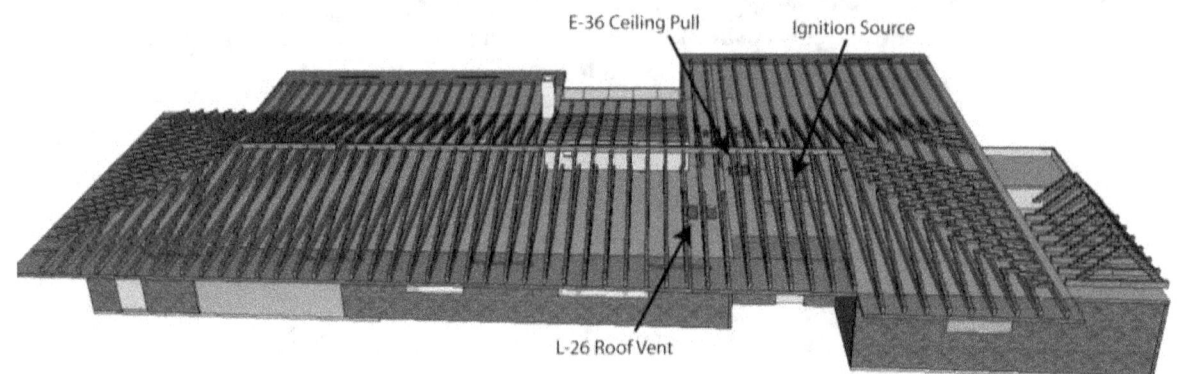

Figure 15: View of the simulated attic construction with rafters visible. The area of fire origin is shown in red. Areas in dark blue are the vent points created by HFD. Image generated using

Based on the timeline provided in Table 1, a minimum of 6 minutes elapsed between when the occupants observed flames in the closet and when E-26 forced open the front door. There are numerous possibilities for how the fire may have spread to the den during this period. Instead of assuming a means of fire spread, the ignition times[6] of the two couches in the den were delayed so that their combined burning produced thermal and fire conditions that resemble the "intense heat" descriptions provided by HFD

---

[6] Ignition time is referred to time zero on the HRR curve provided in Appendix A.

members [30]. The ignition of the couch near to the hallway was delayed by 200 seconds. The ignition of the couch near the solarium glass wall was delayed by 340 seconds.

4.5 Wind

To include the wind present during the fire incident in the FDS simulation required additional model inputs, which are described in this section. The simulation domain size was increased to 56.4 m (185.0 ft) long by 37.2 m (122.0 ft) wide by 11.1 m (36.4 ft) high. The boundaries of the domain were spaced away from the exterior walls of the structure by one length of the structure in each respective axis. The extra distance was provided to resolve wind flow around and over the structure. Increasing the size of the computational domain required additional computational resources. Twenty-five computational meshes were employed, 15 more computational meshes than used in the simulation without wind. Each mesh is shown as a rectangular box in the computational domain shown in Figure 16.

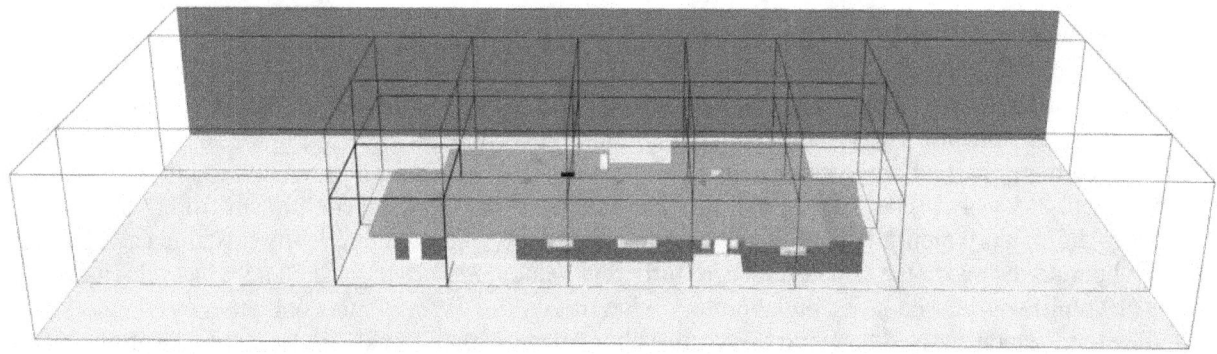

Figure 16: Computational domain of fire simulation with wind

Two computational cell sizes were specified to provide the physical distance required to resolve flow around the structure without increasing computational demands beyond the limits of the computational resources available. A mesh size of 30 cm (1 ft) was used for the larger meshes that "encased" a smaller volume containing the structure. A mesh size of 10 cm (3.9 in) was retained for that volume. The second layer of 10 cm (3.9 in) meshes above the structure was required to provide a resolution fine enough for numerically stable calculation of the higher velocity flow over the surface of the roof. Figure 17 shows the arrangement where the two different mesh sizes abutted. The computational domain totaled approximately ten million computational cells.

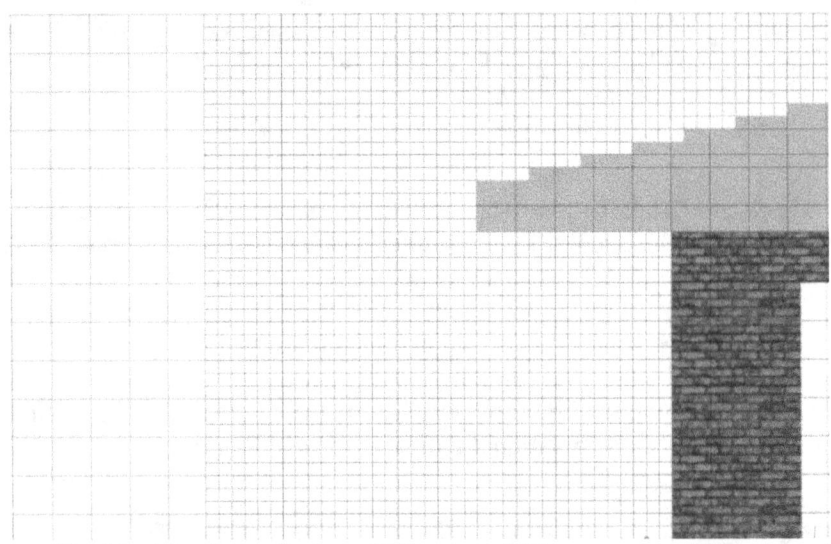
Figure 17: Abutting boundaries of 30 cm and 10 cm computational meshes.

Wind was simulated by applying an orthogonal velocity boundary condition to the eastern face of the computational domain. The velocity boundary condition is shown in blue in Figure 16. Figure 18 shows the trigonometry carried out to estimate the component velocity vector, in turn used to initiate a wind simulation. It was necessary to decrease the final magnitude of the wind component from 7.6 m/s (17 mph) to 4.5 m/s (10 mph). Higher wind velocities frequently caused numerically unstable calculations at mesh boundaries above the structure, where turbulent velocities were highest. The wind velocity of 4.5 m/s (10 mph) was chosen based on laboratory wind driven fire experiments conducted by Madrzykowski and Kerber, which showed the wind driven fire phenomena with external wind velocities as low as 4.5 m/s (10 mph) [8].

Figure 18: Wind vector orthogonal to the computational domain. Background image used with permission of Digital Globe. Enhancements by NIST.

Vents with a specified constant flow rate were applied to simulate wind-driven ventilation provided by the turbine ventilators in the locations shown in Figure 11. A volume flux of 0.175 m^3/s was used for the turbine ventilators, corresponding to an assumed average wind speed of 9.6 m/s (21.5 mph). The volume flux was obtained from experimental data of wind driven turbine ventilators published by Khan [32].

5. Model Results

The results from two different FDS simulations are presented in the following sections: the simulations were conducted with and without wind. The second simulation demonstrates the impact that wind would have on the fire and the fire gas flow path through the structure.

5.1 Structure and Contents without Wind Condition

Starting with the activation of the ignition sources, the simulated fire in the attic spreads from the ignition source to the underside of the roof assembly. The ignition source in the closet ignites the bedding materials on top of the shelves. After 70 seconds, the fire conditions resemble those described by the occupants [11]. Figure 19 shows a view of this initial stage of the fire.

Figure 19: Simulated view of the initial fire in the closet from the kitchen area.

After approximately 5.5 minutes of simulated fire spread, the attic space becomes oxygen-deficient (fuel rich). This under-ventilated condition limits the ignition of materials to near the area of origin and effectively eliminates flaming combustion in the attic until oxygen is later reintroduced by changes in ventilation. Figure 20 shows the oxygen concentration in the attic through the North-South centerline of the house. The area shown in red indicates where the oxygen concentration is less 15 %. Some of the unburned fuels vent from the attic space and burn upon mixing with ambient air, initially from the turbine vent observed by the L-26 captain (Figure 22).

Figure 20: Simulated oxygen concentrations through the centerline of the attic at 5.5 minutes. The plume above the roof in the right hand side shows hot gasses exhausting from a vent.

As the fire develops first in the closet, then in the den, heat and smoke fills the first floor. The thermal and fire conditions that develop in the front hall are similar to those described when the front door was opened [11,12], even with this limited fuel loading.

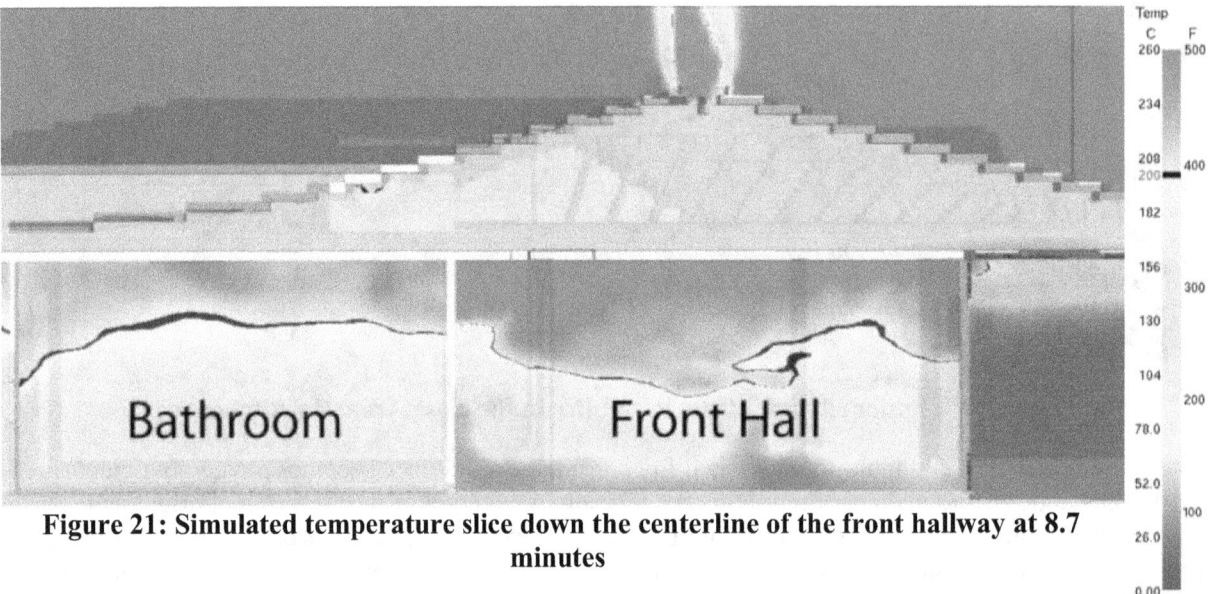

Figure 21: Simulated temperature slice down the centerline of the front hallway at 8.7 minutes

Figure 21 shows the temperatures down the center of the front hall after the front door is opened. Temperatures are approximately 260 °C (500 °F) or above in the upper gas layer, which corresponds with the "extreme heat" felt by HFD members.

In addition to the high temperatures, the first floor is fuel-rich by the time the front door is opened. Flames exit the top of the front door. This was observed by the third member of E-26 while exiting to make a helmet adjustment.

Figure 22: Simulated flames exiting overhead through the front doorway at 8.7 minutes

After the door is opened, fresh outside air flows down the hallway toward the den. Limited flaming combustion resumes in the attic near the area of origin. The ceiling vents, as well as the roof vent and first solarium panel failure have little impact on the fire conditions in the attic or on the first floor.

After 14.3 minutes of simulation time, the remaining solarium glass panels are failed. The flow of the fire gasses is buoyancy-driven. Heat and smoke escape out the failed solarium area as well as into the attic through the failed ceiling area (Figure 23). Contrary to the experience of HFD, temperatures are reduced in the front hallway because less heat and smoke flow into the hallway, and outside air flows into the hallway through the front door. Figure 24 shows the simulated gas temperatures at 1.5 m (5 ft) above the floor throughout the house after the solarium failure. The temperature profile throughout the house shows that at this height, temperatures are highest in the den and closet.

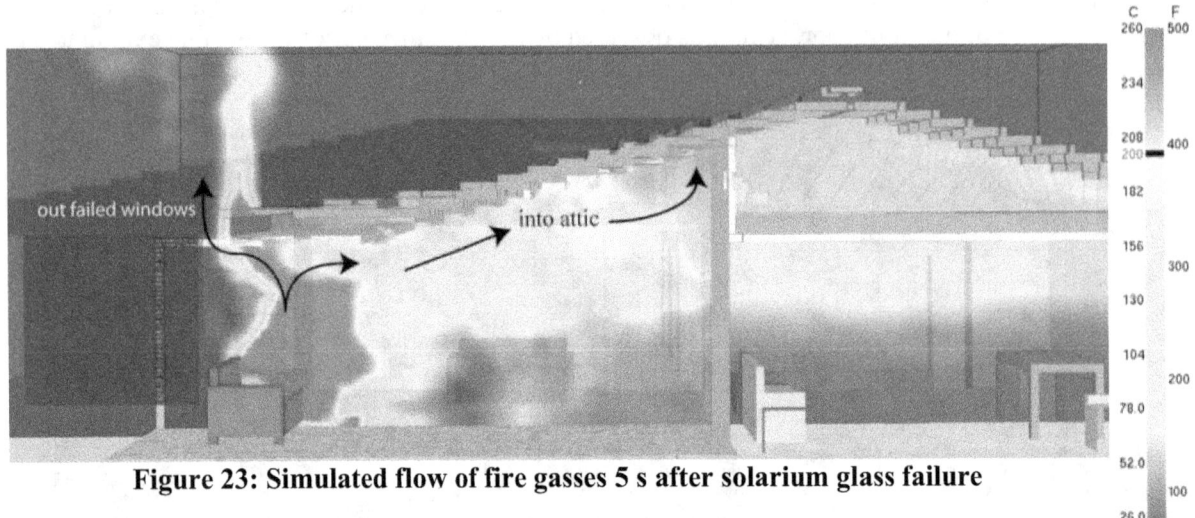

Figure 23: Simulated flow of fire gasses 5 s after solarium glass failure

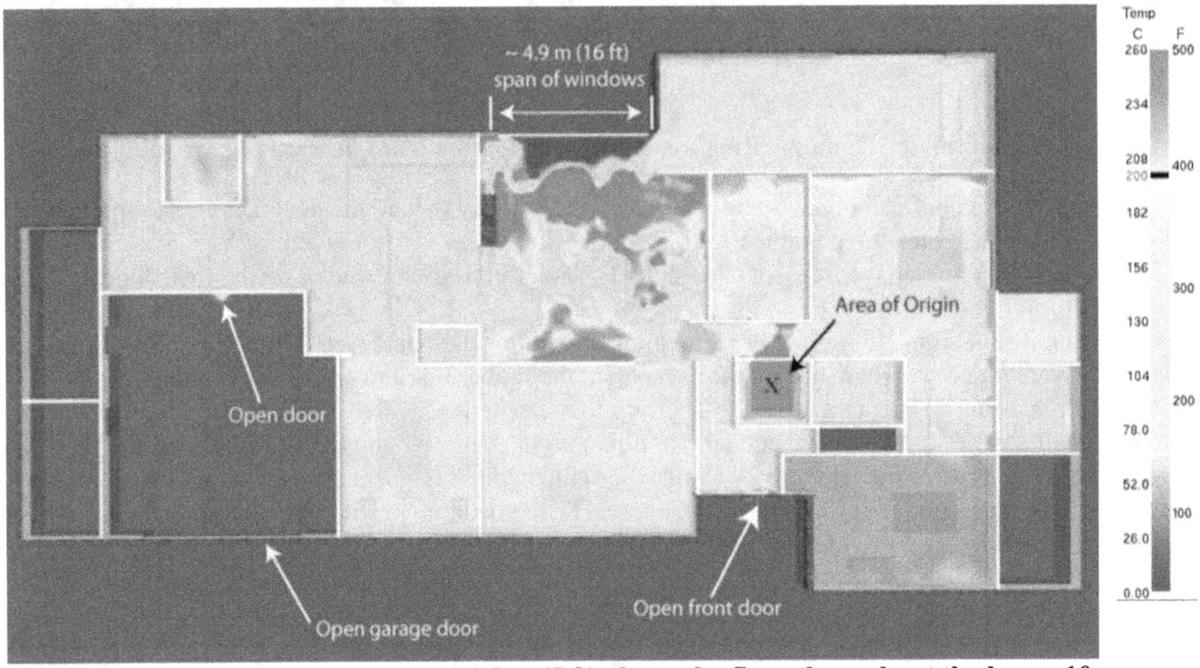

Figure 24: Simulated temperatures at 1.5 m (5 ft) above the floor throughout the house 10 s after solarium failure

5.2 Structure and Contents with Wind Condition

The results of the simulation with wind are largely the same as the results of the simulation without wind until the solarium windows fail after 14.3 minutes of simulation time. Figure 25 demonstrates how the scene may have looked from the street after the occupants exited the house.

Figure 25: Simulated view of the house from the street, demonstrating the visual obscuration challenging arriving apparatus

The attic space is never completely under-ventilated in the wind simulation, whereas the attic was under-ventilated in approximately 5.5 minutes in the no-wind simulation. As a result, burning is continuous in the attic and the fire spreads further through the attic space for the simulation with wind. Figure 26 shows the temperature conditions in the front hallway and attic. Compared with the no-wind simulation, temperatures are approximately the same in the front hallway and higher in the attic.

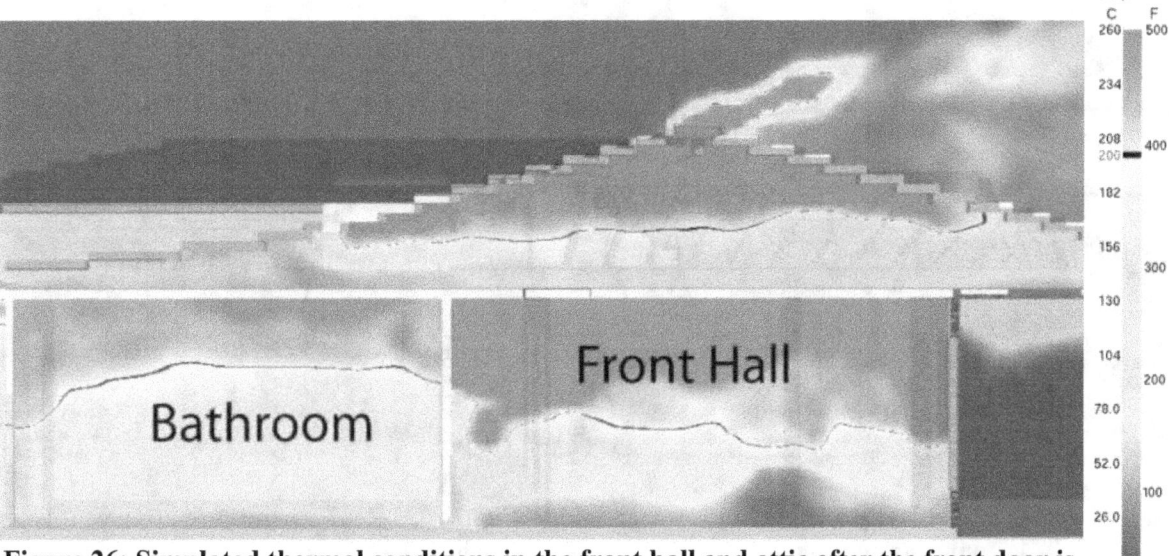

Figure 26: Simulated thermal conditions in the front hall and attic after the front door is opened

More oxygen is available in the attic because the wind provides forced ventilation; wind forces fresh air into the upwind soffit vent on the left, and pressurizes the attic (Figure 27). Fire gasses are forced out the turbine vents and soffit vents (not shown) on the right. When the roof is vented in the wind simulation, fire immediately exits from the vent hole as observed on the fireground. Figure 28 and Figure 29 compare the fire behavior after the roof is vented in the wind simulation and no-wind simulation, respectively.

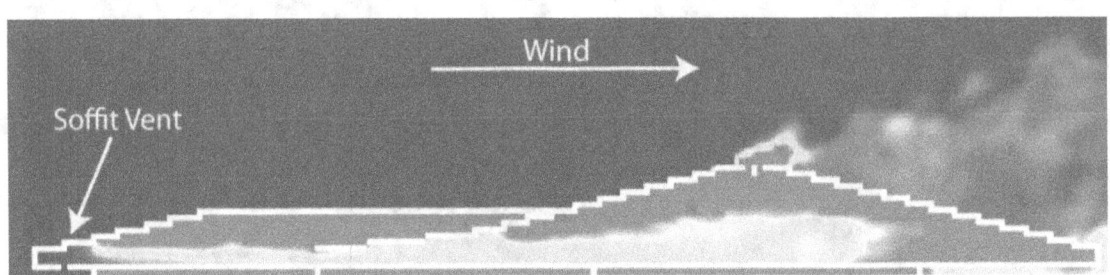

Figure 27: Simulated ventilation conditions in attic at 9.3 minutes

Figure 28: Roof immediately after vent is opened in wind simulation

Figure 29: Roof one minute after vent is opened, when first fire is seen in no-wind simulation

The thermal and fire conditions change dramatically in the structure once the solarium windows fail, due to the wind condition. In Figure 30, just before the solarium windows fail in the wind simulation, the thermal conditions in the front hallway are roughly the same as the conditions at the time the front door is opened. Figure 31 shows the thermal conditions in the front hallway just after the solarium glass fails. In less than 30 s, the temperature in the lower layer transitions from near ambient air temperature to temperatures exceeding 100 °C (212 °F), and the hot gas layer moves closer to the floor. This type of thermal behavior may have been the cause of the intense heat that forced the E-36 crew to the floor while operating in the hall.

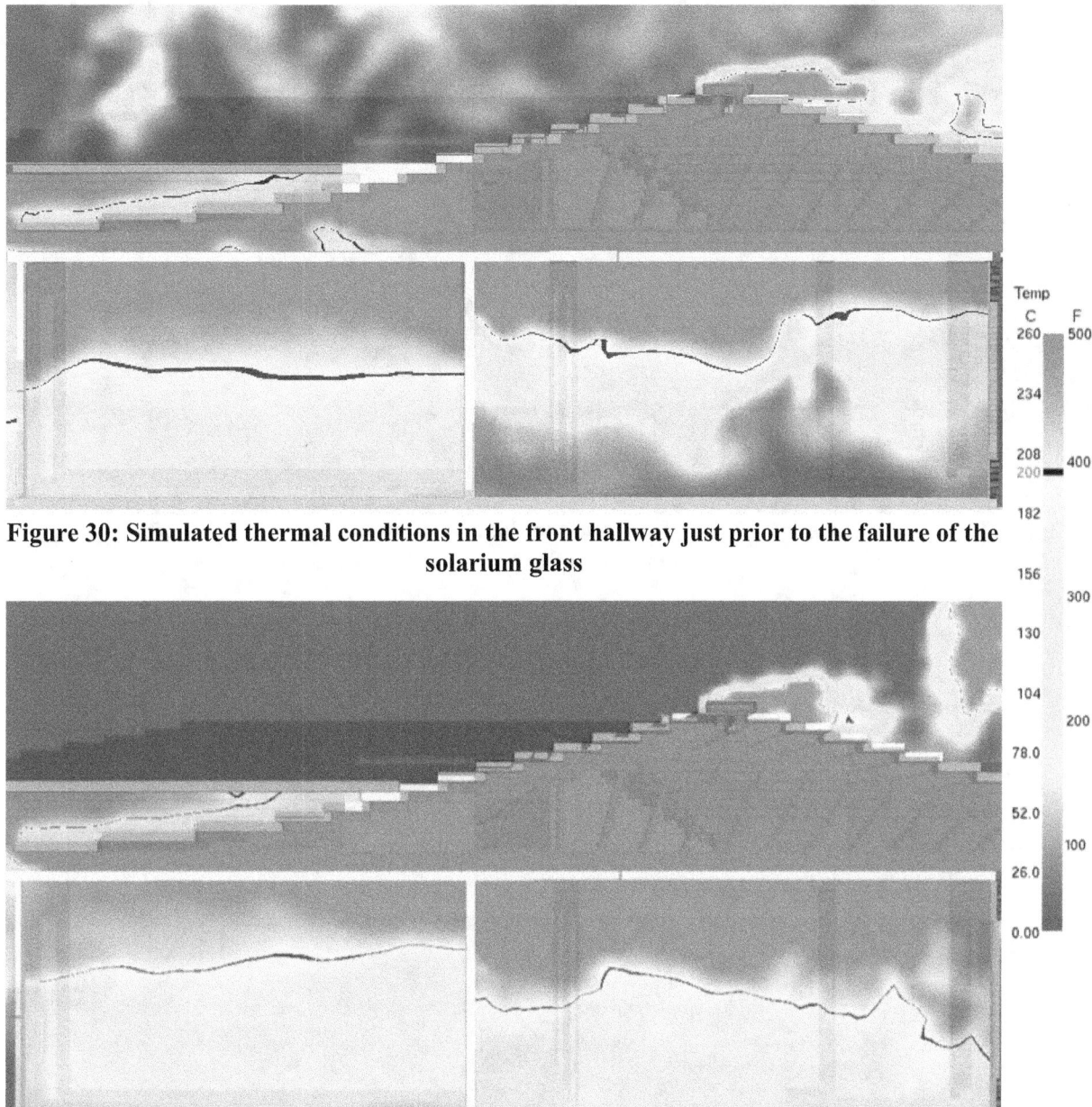

Figure 30: Simulated thermal conditions in the front hallway just prior to the failure of the solarium glass

Figure 31: Simulated thermal conditions in the front hallway just after the failure of the solarium glass

Figure 32 shows the impact of the solarium window failure on the temperature and flow of hot gasses throughout the house at 1.5 m (5ft) above the floor. When the wall of windows fails in the solarium, wind drives ambient air into the den and immediately intensifies the burning of the materials in the area. Figure 32 also shows the flow path of hot gasses driven by wind from side C of the house to side A of the house. Two flow paths are created. Starting in the den, hot gasses travel along the path of least resistance down the hall and out the open front door, as well as through the kitchen and out the open garage door.

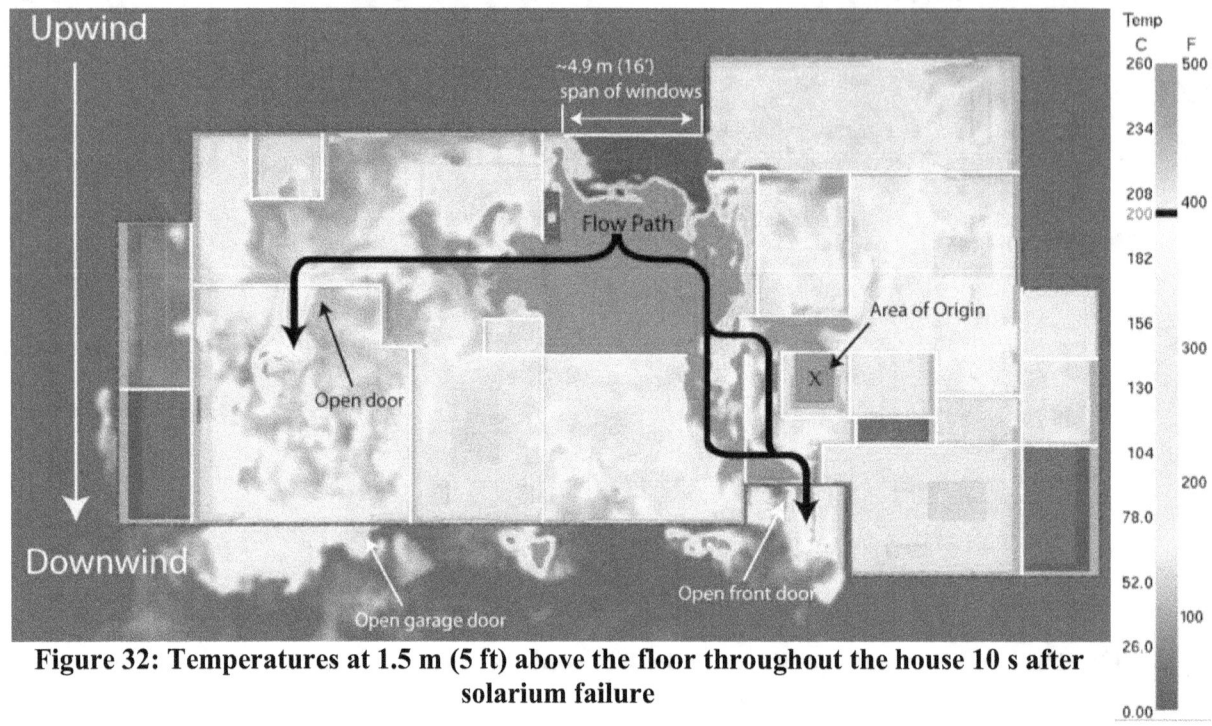

Figure 32: Temperatures at 1.5 m (5 ft) above the floor throughout the house 10 s after solarium failure

6. Discussion

The discussion is divided into two sections. The first section compares the two simulations and makes comparisons with the post fire condition of the structure. The second section addresses tactical considerations based on the information gained from this analysis as well as previous NIST full-scale experiments with and without wind [8,9].

6.1 Comparison of Simulation Results

The wind-driven fire experiments, clearly showed the importance of understanding the ventilation flow path for a fire [8]. A flow path is composed of at least one inlet opening, one exhaust opening, and the connecting volume in between the openings. The direction of the flow is determined by differences in pressure. Heat and smoke in a high pressure area will flow toward areas of lower pressure. The experiments consistently showed that being in the flow path downwind of the fire was only tenable for a fire fighter with full protective clothing and equipment for a limited time, estimated to be 30 s or less. In other words, being in a position between the fire and the exhaust opening or vent, even in non-wind-driven, post-flashover conditions is not a position a fire fighter, or anyone else, should be in.

Figure 24 shows the non-wind-driven fire simulation after the solarium windows failed. The flow path extended from the open front doorway to the fires in the den and the attic. From the fire, most of the heat flowed up and out either through the exhaust opening in the solarium or the exhaust opening up through the attic space. Some of the heat and smoke flowed out of the upper portion of the open front door. In this case, the front doorway serves two functions: the lower portion of the door way is an inlet opening for fresh air, and the upper portion of the doorway is the exhaust for a portion of the fire gasses. The dividing line between the high pressure, hot gas exhaust and the lower pressure cool air inflow is called the neutral plane.

Figure 31 and Figure 32 show the wind-driven fire simulation after the solarium windows failed. The main inlet opening for the flow path is now the solarium windows on the upwind side of the house. The flow interacts with the body of the fire in the den and the attic, and the flow path splits with exhaust up and out the solarium and attic space as before; in addition, the wind forces fire gasses though the house to the exhaust openings at the front doorway and the garage doorway. This change in conditions caused by the wind may explain the "whooshing" sounds and increasing temperatures experienced by E-36 and the conditions that disoriented and overwhelmed the E-26 crew.

Figure 33 shows the increased simulated temperatures in the front hall 5 s after the failure of the solarium windows in the wind-driven case compared to the non-wind-driven case. In the non-wind-driven case, the higher heat conditions (greater than 100 °C (212 °F)) in the gas layer only exhaust out of the top quarter of the doorway opening. In the wind-driven case, the high heat condition exists from the ceiling down to the floor of the front hall, and hot gasses flow out of the front door, top to bottom. Cool air no longer flows into the front doorway. After the failure of the solarium glass, the high pressure from the wind began pushing the fire gasses through the structure from the rear to the front. The front doorway became an exhaust vent, as did the garage door.

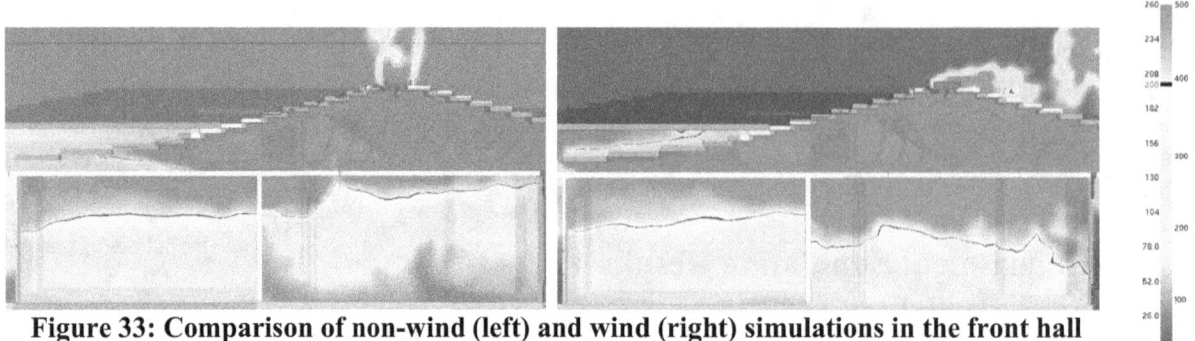

Figure 33: Comparison of non-wind (left) and wind (right) simulations in the front hall area, 5 s after the failure of the solarium glass.

Figure 34 presents the flow vectors at the 1.5 m (5 ft) position above the floor in front doorway and the front hall, looking down at a plan view of the structure. Figure 34 shows the condition 10 s before the failure of the solarium glass (no-wind effect). At this time, cool air is being drawn in to the structure through the front doorway. After the solarium glass failed, the wind caused a flow reversal at the front doorway. The wind entering the structure through the broken solarium glass, increased the pressure at the rear of the structure, and forced the fire gasses through the den into the front hall and out the front door. The same was true for the flow through the kitchen and out through the garage door. It should also be considered that the simulation employed a wind velocity slower than the real wind velocity during the incident. Increased wind velocity would have increased the strength of the flow path.

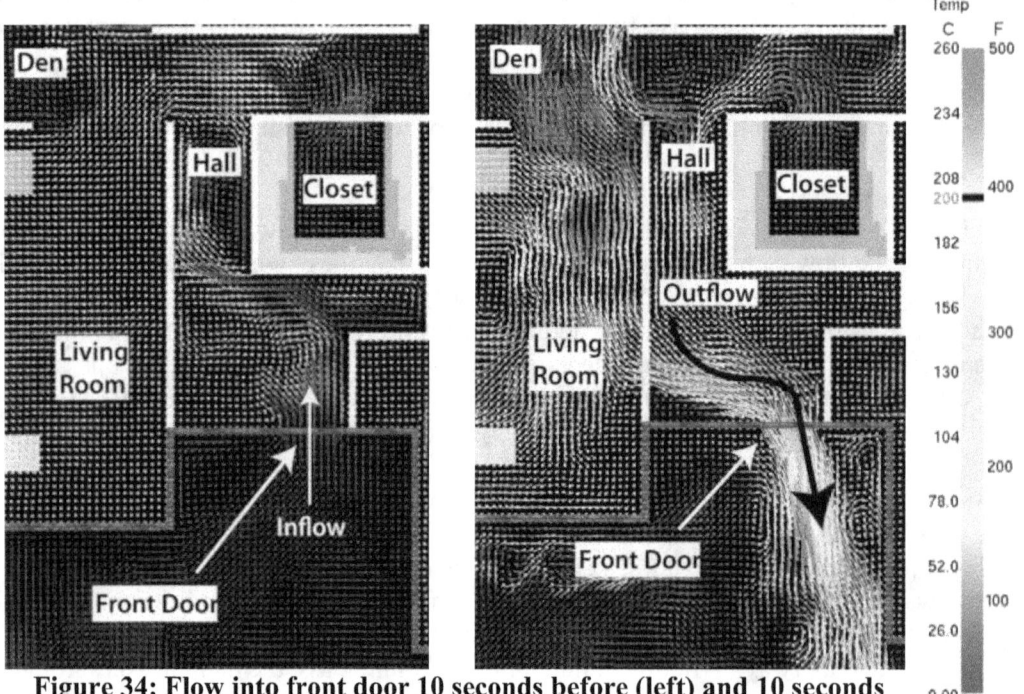

Figure 34: Flow into front door 10 seconds before (left) and 10 seconds after (right) the failure of solarium glass at a plane 1.5 m above the floor

Table 5 provides insight into the change in thermal conditions along the flow path by comparing the average temperatures in each area along the flow path 10 s before the solarium windows failed to the average temperatures 10 s after the solarium windows failed in the wind-driven simulation. The elevation 1.5 m (5 ft) above the floor was chosen for comparison. In addition, the conditions in the living room and dining room are presented. In the den, the temperatures decreased due to the flux of fresh air; however, temperatures in the areas downwind increased. In data collected from full-scale laboratory and field experiments, downwind temperatures have been shown to increase to an even greater extent than shown

36

with the wind-driven simulation [8,9]. The heat losses and the split flow paths may have provided the time to enable the fire fighters closest to the front door to escape. Notice that the highest average temperatures are in the flow path, which includes the den, kitchen and hallway. The lower average temperatures are found in the living room and the dining room. These rooms, while downwind, are not directly in the flow path. However, the temperatures were still too high for a fire fighter in a full structural fire fighting protective ensemble to be exposed to without injury or death. As noted in the *Emergency First Responder Respirator Thermal Characteristics Workshop Proceedings*, the polycarbonate in fire fighter self contained breathing apparatus begins to soften when the material temperature reaches approximately 140°C (284 °F) [33]. Even though fire fighter structural fire fighting coats and pants are tested to withstand temperatures of 260°C (500 °F) [34], the fire fighter inside is susceptible to second degree burn injuries when the inside temperature of the gear exceeds 55°C (130 °F) [35].

Table 5: Comparison of simulated temperatures downwind from the solarium, at 1.5 m (5 ft) above the floor, over the 20 s period beginning 10 s before failure of the solarium glass and ending 10 s after the glass failure.

Location	T_{ave} @5ft, 10 s before solarium failure °C (°F)	T_{ave} @5ft, 10 s after solarium failure °C (°F)	Temperature change over 20s °C (°F)
Den	410 (770)	350 (662)	-60 (-108)
Kitchen	175 (347)	220 (428)	45 (81)
Hallway	175 (347)	220 (428)	45 (81)
Living Room	170 (338)	180 (356)	10 (18)
Dining Room	140 (284)	170 (338)	30 (54)

Figure 35 through Figure 38 are used to show the sections of the structure impacted by the flow path in both the simulation and with photographs taken at the scene. Figure 35 shows the flow path from the upwind side of the house, through the broken solarium windows, and into the den where the flow path splits with some of the heat flowing through the hall to the front doorway, while another portion of the flow passed through the kitchen and into the garage. Black lines and arrows indicate the split flow path through the structure. The temperatures in the figure represent temperatures taken from the simulation at 1.5 m (5 ft) elevation above the floor.

Figure 36 shows the portion of the structure that was involved in the flow path within seconds of the solarium glass failure. The figure contains lines and arrows representing the flow path overlaid on the floor plan with a camera view perspective that shows the area of the structure that is shown in the composite photograph in Figure 37. The photograph shows the level of damage along the flow path, and Figure 38 is included to assist the reader in visualizing the area of the den and the areas leading to the exhaust openings at the front doorway and the doorway to the garage. Comparing Figure 35 with Figure 37, the areas showing simulated temperatures in excess of 260°C (500 °F) have sustained the worst damage, while areas such as the living room and dining room which are at a lower temperature at this time in the fire have less damage.

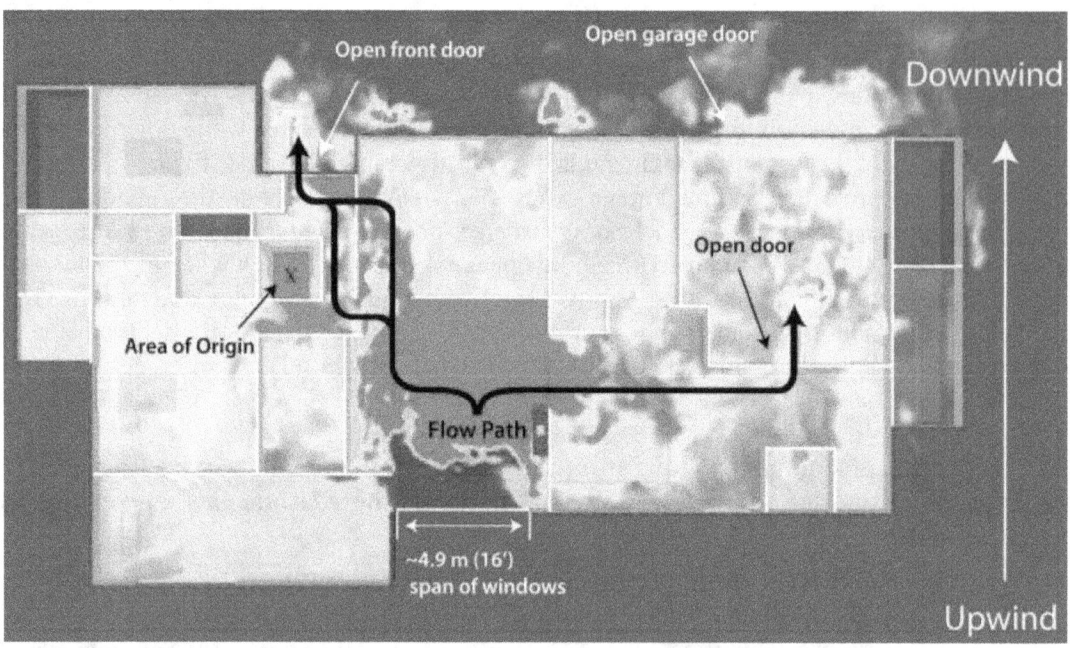

Figure 35: Wind-driven simulation showing the temperatures at 1.5 m (5 ft) above the floor with lines highlighting the direction of the flow path.

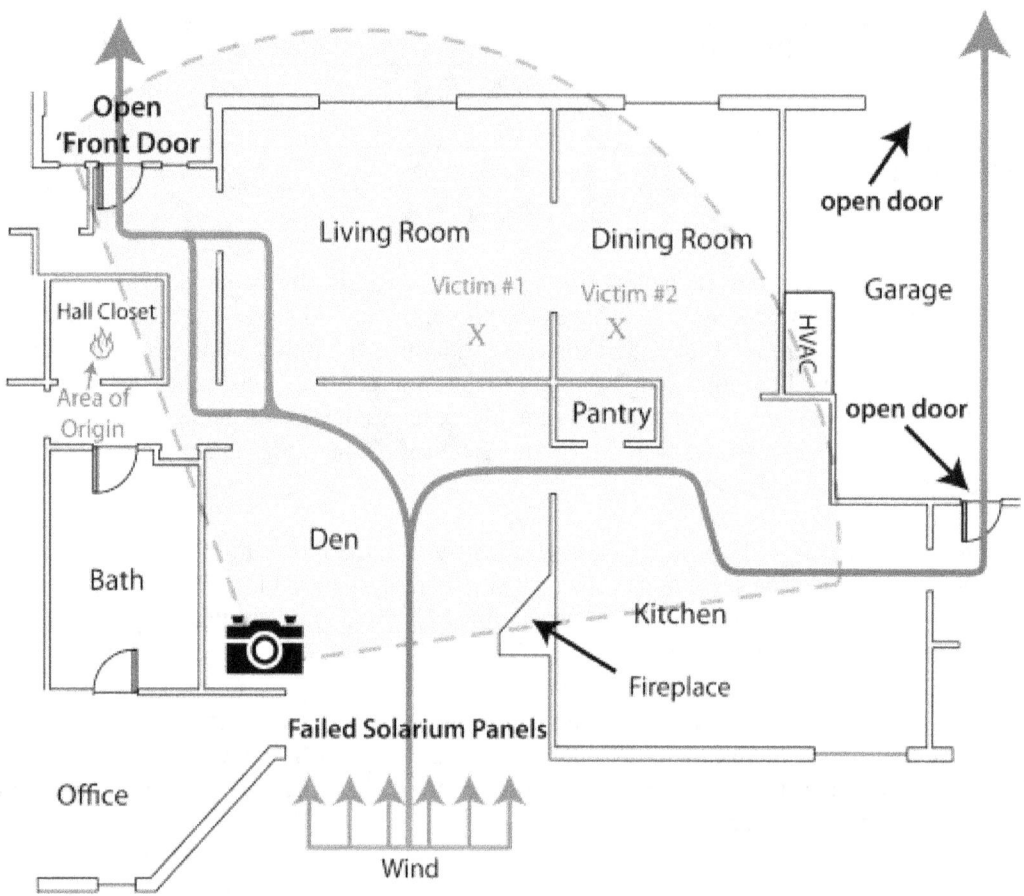

Figure 36: Flow path laid over the floor plan

Figure 37: Flow path laid over panoramic photo taken from the rear of the structure

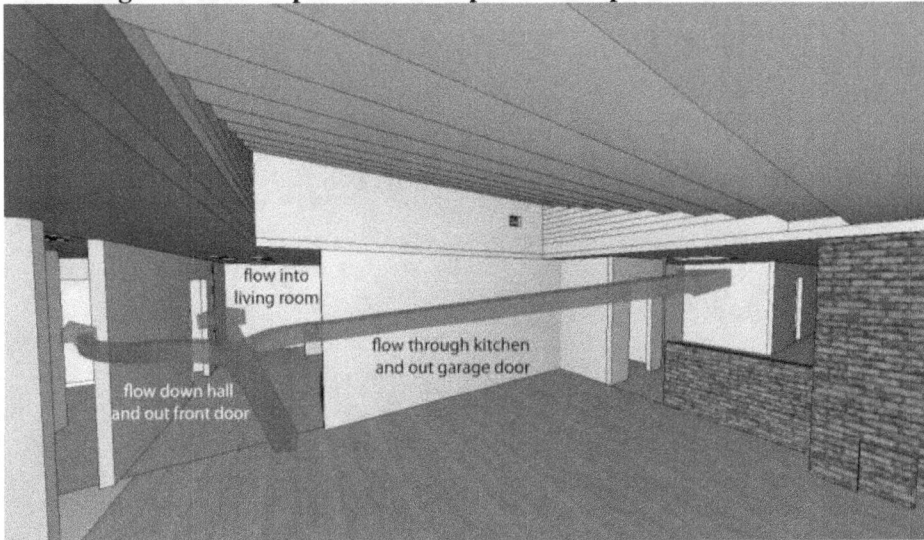

Figure 38: Flow path laid over simulated view of structure as seen from the den. Image generated using

6.2   Tactical Considerations

In these simulations and in previous full scale experiments [8,9], it has been demonstrated that wind can increase the thermal hazards of a structure fire. Therefore, wind needs to be considered as part of the initial "size-up" of the fire conditions and needs to be monitored and reported throughout the fire incident. It is critical for fire fighters to not be in the exhaust portion of the fire flow path. The directional nature of the fire gas flow path results in higher temperatures than the area adjacent to the flow path or upwind of the fire. The flow path can be controlled by limiting ventilation. In this incident, if the front and the garage door were never opened, the only direction for the hot gasses to travel was up through the vents in the roof. Previous studies [8,9] demonstrated that applying water from the exterior, into the upwind side of the structure can have a significant impact on controlling the fire prior to beginning interior operations. Current fire control training guides state, "Whenever possible, approach and attack the fire from the unburned side to keep it from spreading throughout the structure." [36,37]. It should be made clear that in a wind-driven fire, it is most important to use the wind to your advantage and attack the fire from the upwind side of the structure, especially if the upwind side is the burned side. The unexpected ventilation from a broken window can suddenly change the interior thermal conditions. Interior operations need to be aware of potentially rapidly changing conditions.

7. Summary

Two simulations of a fire incident were completed with FDS/Smokeview to provide insight into the fire dynamics in a ranch-style residence that resulted in the fatalities of two Houston fire fighters. The simulations were based on a fire that started with an ignition source on the living space and attic side of the ceiling in a closet near the master bedroom. One simulation included the wind present during the actual incident, and the other did not. Both simulations recreated the witnessed fire behavior and thermal environment until a large span of windows failed on the upwind side of the residence.

In the simulation without wind, the majority of the heat and smoke generated inside the structure exhausted out of the failed span of windows and out of the attic vents. In the no-wind simulation, the end points of the flow path consisted of the air inlet at the front door and hot gas exhaust out the top of the solarium. Therefore, the temperatures in the area between the front door and the den were reduced. These results were not consistent with the reported fire behavior or the resulting damage to the structure.

In the simulation with wind, significant changes in the direction of the hot gas flow occurred after the glass panels in the solarium failed and a strong split flow path was generated. Beginning with the wind inlet at the open solarium wall, the air mixed with pre-heated fuel gasses in the den. Then the flow path split, moving downwind though the structure toward the exhaust openings at the front door and the garage door. Along these flow paths, floor-to-ceiling temperatures increased. This rapid change in the flow path agrees with the rapid increase in thermal exposure experienced by HFD members operating in the downwind path of the wind driven fire gasses. The rapid increase in thermal exposure along the flow paths is what drove interior companies out of the building. and is likely what disoriented and ultimately overwhelmed the two victims.

Wind has been recognized as a contributing factor to fire spread in wildland fires and large-area conflagrations and wildland fire fighters are trained to account for the wind in their tactics. While structural fire departments have recognized the impact of wind on fires, in general, the standard operating guidelines for structural fire fighting have not changed to address the hazards created by a wind driven fire inside a structure. The results of the "no-wind" and the "wind" fire simulations demonstrate how wind conditions can rapidly change the thermal environment from tenable to untenable for fire fighters working in a single-story residential structure fire. The simulation results emphasize the importance of including wind conditions in the scene size-up before beginning and while performing fire fighting operations and adjusting tactics based on the wind conditions. These results are in agreement with NIST studies conducted to examine wind driven fire conditions in high-rise structures.

Based on the analysis of this fire incident and results from previous studies, adjusting fire fighting tactics to account for wind conditions in structural fire fighting is critical to enhance the safety and the effectiveness of fire fighters.

8. Acknowledgements

The authors would like to thank the Houston Fire Department, especially Executive Assistant Chief Carl Matejka and District Chief Michael Shrum (ret.). The authors also thank Tim Mariner of the NIOSH Fire Fighter Fatality Investigation and Prevention Program for support of this research. The authors extend their appreciation to Special Agent Brian M. Hoback of the Bureau of Alcohol, Tobacco and Firearms for thorough documentation of the structure involved in this fire incident. The authors also thank Roy McLane, Dave Stacy and Chris Leier of NIST. Finally, we would like to thank Kevin McGrattan for leading the continued development of the NIST Fire Dynamics Simulator Model and Glenn Forney for development of Smokeview.

42

9. References

[1] D. Madrzykowski and R. Vettori, "Simulation of the Dynamics of the Fire at 3146 Cherry Road NE Washington D.C., May 30, 1999," National Institute of Standards and Technology, Gaithersburg, MD, NISTIR 6510 2000.

[2] D. Madrzykowski, G. Forney, and D. Walton, "Simulation of the Dynamics of a Fire in a Two-Story Duplex - Iowa, December 22, 1999," National Institute of Standards and Technology, Gaithersburg, MD, NISTIR 6854 2002.

[3] R. Vettori, D. Madrzykowski, and W. Walton, "Simulation of the Dynamics of a Fire in a One-Story Restaurant -- Texas, February 14, 2000," National Institute of Standards and Technology, Gaithersburg, MD, NISTIR 6923 2002.

[4] D. Madrzykowski and W. Walton, "Cook County Administration Building Fire, 69 West Washington, Chicago, Illinois, October 17, 2003: Heat Release Rate Experiments and FDS Simulations," National Institute of Standards and Technology, Gaithersburg, MD, NIST SP 1021 2004.

[5] K. McGrattan, C. Bouldin, and G. Forney, "Computer Simulation of the Fires in the World Trade Center Towers. Federal Building and Fire Safety Investigation of the World Trade Center Disaster.," National Institute of Standards and Technology, Gaithersburg, MD, NIST NCSTAR 1-5F 2005.

[6] W. Grosshandler, N. Bryner, D. Madrzykowski, and K. Kuntz, "Report of the Technical Investigation of The Station Nightclub Fire," National Institute of Standards and Technology, Gaithersburg, MD, NIST NCSTAR 2 2005.

[7] N. Bryner, P. Fuss, B. Klein, and A. Putorti, "Technical Study of the Sofa Super Store Fire - South Carolina, June 18, 2007," National Institute of Standards and Technology, Gaithersburg, MD, 2011.

[8] D. Madrzykowski and S. Kerber, "Fire Fighting Tactics Under Wind Driven Conditions: Laboratory Experiments," National Institute of Standards and Technology, Gaithersburg, MD, TN 1618 January 2009.

[9] S. Kerber and D. Madrzykowski, "Fire Fighting Tactics Under Wind Driven Fire Conditions: 7-Story Building Experiments," National Institute of Standards and Technology, Gaithersburg, MD, TN 1629 April 2009.

[10] National Institute for Occupational Safety and Health, "Career Fire Fighter Dies in Wind Driven Residential Structure Fire - Virginia," National Institute for Occupational Safety and Health, Morgantown, WV, 2008.

[11] Texas State Fire Marshal's Office, "State Fire Marshal's Office Fire Fighter Fatality Investigation Case FY09-01," Texas Department of Insurance, Austin, Texas, 2009.

[12] The National Institute for Occupational Safety and Health, "Career Probationary Fire Fighter and Captain Die as a Result of Rapid Fire Progression in a Wind-Driven Residential Structure Fire - Texas," The National Institute for Occupational Safety and Health, Cincinatti, OH, April 8, 2010.

[13] K. McGrattan, R., Hostikka, S. McDermott, and J. Floyd, "Fire Dynamics Simulator (Version 5) User's Guide," National Institute of Standards and Technology, Gaithersburg, MD, NIST SP 1019-5 October 29, 2010.

[14] D. Drysdale, *An Introduction to Fire Dynamics*, 2nd ed. West Sussex, England: John Wiley & Sons, 1998.

[15] K. McGrattan, C. Bouldin, and G. Forney, "Computer Simulation of the Fires in the World Trade Center Towers," National Institute of Standards and Technology, Gaithersburg, MD, U.S.A., NCSTAR 1-5F September 2005.

[16] T.K. Chan and D.H. Napier, "Smoldering and Ignition of Cotton Fibres and Dust"," *Fire Prevention Science and Technology*, no. 4, pp. 13-23, February 1973.

[17] National Institute of Standards and Technology. (2011, April) Test Type: Cone Calorimeter, Test Number: t5603, Description File: t5603. [Online]. http://www.fire.nist.gov/fastdata/data/t5603/t5603.htm

[18] J. Watanabe and T. Takeyoshi, "Experimental Investigation into Penetration of a Weak Fire Plume into a Hot Upper Layer," *Journal of Fire Sciences*, vol. 22, pp. 405-420, September 2004.

[19] Ogden Manufacturing Company. (2011, April) Engineering & Technical. [Online]. http://www.ogdenmfg.com/pdf/tech9.pdf

[20] M. Spearpoint, "Predicting the Ignition and Burning of Wood in the Cone Calorimeter Using an Integral Model," National Institute of Standards and Technology, Gaithersburg, MD, GCR 99-775 1999.

[21] A. Tewarson, "Generation of Heat and Gaseous, Liquid, and Solid Products in Fires," in *The SFPE Handbook of Fire Protection Engineering*. Bethesda, MD, U.S.A.: National Fire Protection Association, 2008, ch. 3-3, pp. 3-142.

[22] Forest Products Laboratory., "Chapter 04: Moisture Relations and Physical Properties of Wood," in *Centennial edition of Wood handbook, Wood as an Engineering Material*. Madison, WI, USA: Forest Products Laboratory, 2010, p. 508.

[23] Engineering Toolbox. (2011, April) The Engineering Toolbox. [Online]. http://www.engineeringtoolbox.com/thermal-conductivity-d_429.html

[24] United States Department of Energy. (2011, September) Energy Savers: Loose-Fill Insulation. [Online]. http://www.energysavers.gov/your_home/insulation_airsealing/index.cfm/mytopic=11650

[25] G. Wilkes, *Heat Insulation*. U.S.: John Wiley & Sons, 1950.

[26] B.H. Jones, "Performance of Gypsum Plasterboard Assemblies Exposed to Real Building Fires," University of Canterbury, Christchurch, NZ, 2001.

[27] National Institute of Standards and Technology. (1990, December) Test Type: Cone Calorimeter, Test Number: 11, Description File: f206. [Online]. http://fire.nist.gov/fastdata/docs/sp199124.htm

[28] International Code Council, *International Residential Code for One- and Two-Family Dwellings*. Country Club Hills, IL, U.S.A.: International Code Council, Inc., 2006.

[29] R. Dumont, H. Orr, and M. Lux, "Low Energy Prarie Housing," National Research Council of Canada, Ottowa, Building Practice Note 1982.

[30] Houston Arson Bureau, Voluntary Statements of Houston Fire Department Members to Houston Arson Bureau, 2009.

[31] J. Shanley, "USFA Fire Burn Pattern Tests," United States Fire Administration, 1997.

[32] N. Khan, Y Su, S Riffat, and C. Biggs, "Performance testing and comparison of turbine ventilators," *Renewable Energy*, no. 33, pp. 2441-2447, January 2008.

[33] A. Mensch and N. Bryner, "Emergency First Responder Respirator Thermal Characteristics Workshop Proceedings," National Institute of Standards and Technology, Gaithersburg, MD, NIST SP 1123 June 2011.

[34] National Fire Protection Association, "NFPA 1971: Standard on Protective Ensembles for Structural Fire Fighting and Proximity Fire Fighting," National Fire Protection Association, Quincy, MA, 2007.

[35] American Society for Testing and Materials, "C 1055 Standard Guide for Heated Systems Surface Conditions that Produce Contact Burn Injuries," American Society for Testing and Materials, West Conshohocken, PA, 2009.

[36] The International Fire Service Training Association, *Essentials of Fire Fighting*, 5th ed., Carl Goodson and Lynne Murnane, Eds. Stillwater, OK, USA: Fire Protection Publications, 2008.

[37] Thomson Delmar Learning, *Firefighter's Handbook*, 2nd ed. Clifton Park, NY, USA: Thomson Delmar Learning, Essentials of Firefighting and Emergency Response.

[38] K. R. Prasad, R. Kramer, N. Marsh, and M. R. Nyden, "Numerical Simulation of Fire Spread on Polyurethane Foam Slabs.," in *Fire and Materials 11th International Conference.*, San Francisco, CA, January 26-28, 2009, pp. 697-708.

[39] The Engineering Toolbox. (2011, April) Engineering Toolbox. [Online]. http://www.engineeringtoolbox.com/fiberglas-insulation-k-values-d_1172.html

[40] T. Ohlemiller, J. Shields, R. McLane, and R. Gann, "Flammability Assessment Methods for Mattresses," National Institute of Standards and Technology, Gaithersburg, MD, NISTIR 6497 June 2000.

[41] R. Bilbao, J. Mastral, J. Ceamanos, and M. Aldea, "Kinetics of the Thermal Decomposition of Polyurethane Foams in Nitrogen and Air Atmospheres," *Journal of Analytical and Applied Pyrolysis*, no. 37, pp. 69-82, February 1996.

[42] M. Graham, "Principles of attic ventilation," *Professional Roofing Magazine*, January 2001.

[43] M. K. Donnelly, W. D. Davis, J. R. Lawson, and M. J. Selepak, "Thermal Environment for Electronic Equipment Used by First Responders," NIST, Gaithersburg, MD, Technical Note 1474, 2006.

[44] D. Madrzykowski and S. Kerber. (2010, March) Wind Driven Fire Research: Hazards and Tactics. [Online]. http://www.fireengineering.com/articles/2010/03/wind-driven-fire-research-hazards-and-tactics.html

Appendix A: FDS Input File

```
&HEAD CHID='WF16slow'/
&TIME T_END=980.00/
&DUMP RENDER_FILE='WF16slow.ge1', DT_PL3D=100.00, NFRAMES=1960/
&MISC TMPA=21.00, BAROCLINIC=.TRUE./
&MESH ID='MESH 1', IJK=78,102,63, XB=-3.40,4.40,-1.90,8.30,-0.1,6.20/
&MESH ID='MESH 2', IJK=84,102,63, XB=4.40,12.80,-1.90,8.30,-0.1,6.20/
&MESH ID='MESH 3', IJK=84,102,63, XB=12.80,21.20,-1.90,8.30,-0.1,6.20/
&MESH ID='MESH 4', IJK=72,102,63, XB=21.20,28.40,-1.90,8.30,-0.1,6.20/
&MESH ID='MESH 5', IJK=72,102,63, XB=28.40,35.60,-1.90,8.30,-0.1,6.20/
&MESH ID='MESH 6', IJK=78,108,63, XB=-3.40,4.40,8.30,19.10,-0.1,6.20/
&MESH ID='MESH 7', IJK=84,108,63, XB=4.40,12.80,8.30,19.10,-0.1,6.20/
&MESH ID='MESH 8', IJK=84,108,63, XB=12.80,21.20,8.30,19.10,-0.1,6.20/
&MESH ID='MESH 9', IJK=72,108,63, XB=21.20,28.40,8.30,19.10,-0.1,6.20/
&MESH ID='MESH10', IJK=72,108,63, XB=28.40,35.60,8.30,19.10,-0.1,6.20/
&MESH ID='MESH 11', IJK=78,102,48, XB=-3.40,4.40,-1.90,8.30,6.20,11.00/
&MESH ID='MESH 12', IJK=84,102,48, XB=4.40,12.80,-1.90,8.30,6.20,11.00/
&MESH ID='MESH 13', IJK=84,102,48, XB=12.80,21.20,-1.90,8.30,6.20,11.00/
&MESH ID='MESH 14', IJK=72,102,48, XB=21.20,28.40,-1.90,8.30,6.20,11.00/
&MESH ID='MESH 15', IJK=72,102,48, XB=28.40,35.60,-1.90,8.30,6.20,11.00/
&MESH ID='MESH 16', IJK=78,108,48, XB=-3.40,4.40,8.30,19.10,6.20,11.00/
&MESH ID='MESH 17', IJK=84,108,48, XB=4.40,12.80,8.30,19.10,6.20,11.00/
&MESH ID='MESH 18', IJK=84,108,48, XB=12.80,21.20,8.30,19.10,6.20,11.00/
&MESH ID='MESH 19', IJK=72,108,48, XB=21.20,28.40,8.30,19.10,6.20,11.00/
&MESH ID='MESH 20', IJK=72,108,48, XB=28.40,35.60,8.30,19.10,6.20,11.00/
&MESH ID='MESHW1', IJK=248,27,37, XB=-21.40,53.00,-10.00,-1.90,-0.1,11.00/
&MESH ID='MESHW2', IJK=248,27,37, XB=-21.40,53.00,19.10,27.20,-0.1,11.00/
&MESH ID='MESHW3', IJK=60,70,37, XB=-21.40,-3.40,-1.90,19.10,-0.1,11.00/
&MESH ID='MESHW4', IJK=58,70,37, XB=35.60,53.00,-1.90,19.10,-0.1,11.00/
&REAC ID='Wood',
      FYI='Ritchie, et al., 5th IAFSS, C_3.4 H_6.2 O_2.5',
      C=1.00,
      H=1.70,
      O=0.74,
      N=2.00E-3,
      HEAT_OF_COMBUSTION=1.30E4/
&MATL ID='Gypsum',
      FYI='Drysdale, P. 36',
      SPECIFIC_HEAT=1.09,
      CONDUCTIVITY=0.1700,
      DENSITY=930.00/
&MATL ID='Fiberglass Insulation',
      FYI='fiberglass r13, values from engineering toolbox, cp is from glass',
      SPECIFIC_HEAT=0.84,
      CONDUCTIVITY=0.0400,
      DENSITY=14.00/
&MATL ID='DougFir',
      FYI='Ref=Spearpoint''s thesis, except cp',
      SPECIFIC_HEAT=1.80,
      CONDUCTIVITY=0.1300,
      DENSITY=502.00,
      HEAT_OF_COMBUSTION=1.30E4/
&MATL ID='Glass',
      FYI='Drysdale, P. 36',
      SPECIFIC_HEAT=0.84,
      CONDUCTIVITY=0.76,
      DENSITY=2.70E3/
&MATL ID='Cotton',
      SPECIFIC_HEAT=1.29,
      CONDUCTIVITY=0.1100,
      DENSITY=16.40,
      HEAT_OF_COMBUSTION=1.56E4/
&MATL ID='Brick',
      FYI='Drysdale, P. 36',
      SPECIFIC_HEAT=0.84,
      CONDUCTIVITY=0.69,
      DENSITY=1.60E3/
&MATL ID='Carpet',
      FYI='WTC report value',
      SPECIFIC_HEAT=4.50,
      CONDUCTIVITY=0.1600,
      DENSITY=750.00,
      HEAT_OF_COMBUSTION=2.23E4/
&SURF ID='Ceiling Gypsum',
      COLOR='TAN',
      HRRPUA=225.00,
      RAMP_Q='Ceiling Gypsum_RAMP_Q',
      IGNITION_TEMPERATURE=400.00,
      BACKING='EXPOSED',
      MATL_ID(1,1)='Gypsum',
      MATL_ID(2,1)='Fiberglass Insulation',
      MATL_MASS_FRACTION(1,1)=1.00,
      MATL_MASS_FRACTION(2,1)=1.00,
      THICKNESS(1:2)=0.0200,0.0900/
&RAMP ID='Ceiling Gypsum_RAMP_Q', T=0.00, F=0.0100/
&RAMP ID='Ceiling Gypsum_RAMP_Q', T=2.00, F=0.0200/
&RAMP ID='Ceiling Gypsum_RAMP_Q', T=3.64, F=0.1600/
&RAMP ID='Ceiling Gypsum_RAMP_Q', T=4.24, F=0.40900/
```

```
&RAMP ID='Ceiling Gypsum_RAMP_Q', T=6.67, F=0.78/
&RAMP ID='Ceiling Gypsum_RAMP_Q', T=7.88, F=1.00/
&RAMP ID='Ceiling Gypsum_RAMP_Q', T=15.76, F=0.40900/
&RAMP ID='Ceiling Gypsum_RAMP_Q', T=18.18, F=0.2600/
&RAMP ID='Ceiling Gypsum_RAMP_Q', T=24.85, F=0.1100/
&RAMP ID='Ceiling Gypsum_RAMP_Q', T=41.21, F=0.0600/
&RAMP ID='Ceiling Gypsum_RAMP_Q', T=62.42, F=0.0700/
&RAMP ID='Ceiling Gypsum_RAMP_Q', T=78.79, F=0.0700/
&SURF ID='Attic Gypsum',
      RGB=204,153,0,
      BACKING='EXPOSED',
      MATL_ID(1,1)='Fiberglass Insulation',
      MATL_ID(2,1)='Gypsum',
      MATL_MASS_FRACTION(1,1)=1.00,
      MATL_MASS_FRACTION(2,1)=1.00,
      THICKNESS(1:2)=0.0900,0.0200/
&SURF ID='Closet Burner',
      COLOR='RED',
      HRRPUA=280.00,
      RAMP_Q='Closet Burner_RAMP_Q'/
&RAMP ID='Closet Burner_RAMP_Q', T=0.00, F=0.00/
&RAMP ID='Closet Burner_RAMP_Q', T=3.00, F=1.00/
&RAMP ID='Closet Burner_RAMP_Q', T=100.00, F=1.00/
&RAMP ID='Closet Burner_RAMP_Q', T=101.00, F=0.00/
&SURF ID='Attic Burner',
      COLOR='RED',
      CONVECTIVE_HEAT_FLUX=75.00,
      RAMP_Q='Attic Burner_RAMP_Q2',
      HRRPUA=1.00E3/
&RAMP ID='Attic Burner_RAMP_Q', T=0.00, F=0.00/
&RAMP ID='Attic Burner_RAMP_Q', T=10.00, F=1.00/
&RAMP ID='Attic Burner_RAMP_Q', T=150.00, F=1.00/
&RAMP ID='Attic Burner_RAMP_Q', T=151.00, F=0.00/
&RAMP ID='Attic Burner_RAMP_Q2', T=0.00, F=0.00/
&RAMP ID='Attic Burner_RAMP_Q2', T=10.00, F=1.00/
&RAMP ID='Attic Burner_RAMP_Q2', T=150.00, F=1.00/
&RAMP ID='Attic Burner_RAMP_Q2', T=151.00, F=0.00/
&SURF ID='Gypsum',
      COLOR='TAN',
      HRRPUA=225.00,
      RAMP_Q='Gypsum_RAMP_Q',
      IGNITION_TEMPERATURE=400.00,
      MATL_ID(1,1)='Gypsum',
      MATL_MASS_FRACTION(1,1)=1.00,
      THICKNESS(1)=0.0127/
&RAMP ID='Gypsum_RAMP_Q', T=0.00, F=0.0100/
&RAMP ID='Gypsum_RAMP_Q', T=2.00, F=0.0200/
&RAMP ID='Gypsum_RAMP_Q', T=3.64, F=0.1600/
&RAMP ID='Gypsum_RAMP_Q', T=4.24, F=0.40900/
&RAMP ID='Gypsum_RAMP_Q', T=6.67, F=0.78/
&RAMP ID='Gypsum_RAMP_Q', T=7.88, F=1.00/
&RAMP ID='Gypsum_RAMP_Q', T=15.76, F=0.40900/
&RAMP ID='Gypsum_RAMP_Q', T=18.18, F=0.2600/
&RAMP ID='Gypsum_RAMP_Q', T=24.85, F=0.1100/
&RAMP ID='Gypsum_RAMP_Q', T=41.21, F=0.0600/
&RAMP ID='Gypsum_RAMP_Q', T=62.42, F=0.0700/
&RAMP ID='Gypsum_RAMP_Q', T=78.79, F=0.0700/
&SURF ID='Closet Walls',
      FYI='Tig from Janssens, cone derived props from Spearpoint',
      RGB=146,202,166,
      TEXTURE_MAP='psm_spruce.jpg',
      TEXTURE_WIDTH=0.67,
      TEXTURE_HEIGHT=2.44,
      HRRPUA=181.39,
      RAMP_Q='Closet Walls_RAMP_Q',
      IGNITION_TEMPERATURE=384.00,
      BACKING='EXPOSED',
      MATL_ID(1,1)='DougFir',
      MATL_ID(2,1)='Gypsum',
      MATL_MASS_FRACTION(1,1)=1.00,
      MATL_MASS_FRACTION(2,1)=1.00,
      THICKNESS(1:2)=6.00E-3,0.0127/
&RAMP ID='Closet Walls_RAMP_Q', T=0.00, F=0.00/
&RAMP ID='Closet Walls_RAMP_Q', T=0.1000, F=0.00/
&RAMP ID='Closet Walls_RAMP_Q', T=0.3800, F=0.3600/
&RAMP ID='Closet Walls_RAMP_Q', T=1.61, F=0.69/
&RAMP ID='Closet Walls_RAMP_Q', T=2.52, F=0.94/
&RAMP ID='Closet Walls_RAMP_Q', T=8.19, F=1.00/
&RAMP ID='Closet Walls_RAMP_Q', T=10.11, F=0.94/
&RAMP ID='Closet Walls_RAMP_Q', T=15.38, F=0.89/
&RAMP ID='Closet Walls_RAMP_Q', T=20.72, F=0.83/
&RAMP ID='Closet Walls_RAMP_Q', T=22.83, F=0.72/
&RAMP ID='Closet Walls_RAMP_Q', T=38.51, F=0.62/
&RAMP ID='Closet Walls_RAMP_Q', T=60.92, F=0.54/
&RAMP ID='Closet Walls_RAMP_Q', T=91.81, F=0.40600/
&RAMP ID='Closet Walls_RAMP_Q', T=127.72, F=0.40/
&RAMP ID='Closet Walls_RAMP_Q', T=171.96, F=0.3900/
&RAMP ID='Closet Walls_RAMP_Q', T=241.78, F=0.3600/
&RAMP ID='Closet Walls_RAMP_Q', T=323.44, F=0.3500/
&RAMP ID='Closet Walls_RAMP_Q', T=376.20, F=0.3300/
&RAMP ID='Closet Walls_RAMP_Q', T=442.49, F=0.3300/
&RAMP ID='Closet Walls_RAMP_Q', T=522.44, F=0.3200/
&RAMP ID='Closet Walls_RAMP_Q', T=602.31, F=0.3300/
&RAMP ID='Closet Walls_RAMP_Q', T=677.15, F=0.3200/
&RAMP ID='Closet Walls_RAMP_Q', T=784.29, F=0.3100/
```

```
&RAMP ID='Closet Walls_RAMP_Q', T=940.64, F=0.3300/
&RAMP ID='Closet Walls_RAMP_Q', T=1.0494200E3, F=0.3400/
&RAMP ID='Closet Walls_RAMP_Q', T=1.1650400E3, F=0.3400/
&RAMP ID='Closet Walls_RAMP_Q', T=1.2398600E3, F=0.3300/
&RAMP ID='Closet Walls_RAMP_Q', T=1.3435500E3, F=0.3300/
&RAMP ID='Closet Walls_RAMP_Q', T=1.4115700E3, F=0.3300/
&SURF ID='Roof Sheathing',
      RGB=153,102,0,
      HRRPUA=181.39,
      RAMP_Q='Roof Sheathing_RAMP_Q',
      IGNITION_TEMPERATURE=384.00,
      BACKING='EXPOSED',
      MATL_ID(1,1)='DougFir',
      MATL_MASS_FRACTION(1,1)=1.00,
      THICKNESS(1)=0.0190/
&RAMP ID='Roof Sheathing_RAMP_Q', T=0.00, F=0.00/
&RAMP ID='Roof Sheathing_RAMP_Q', T=0.1000, F=0.00/
&RAMP ID='Roof Sheathing_RAMP_Q', T=0.3800, F=0.3600/
&RAMP ID='Roof Sheathing_RAMP_Q', T=1.61, F=0.69/
&RAMP ID='Roof Sheathing_RAMP_Q', T=2.52, F=0.94/
&RAMP ID='Roof Sheathing_RAMP_Q', T=8.19, F=1.00/
&RAMP ID='Roof Sheathing_RAMP_Q', T=10.11, F=0.94/
&RAMP ID='Roof Sheathing_RAMP_Q', T=15.38, F=0.89/
&RAMP ID='Roof Sheathing_RAMP_Q', T=20.72, F=0.83/
&RAMP ID='Roof Sheathing_RAMP_Q', T=22.83, F=0.72/
&RAMP ID='Roof Sheathing_RAMP_Q', T=38.51, F=0.62/
&RAMP ID='Roof Sheathing_RAMP_Q', T=60.92, F=0.54/
&RAMP ID='Roof Sheathing_RAMP_Q', T=91.81, F=0.40600/
&RAMP ID='Roof Sheathing_RAMP_Q', T=127.72, F=0.40/
&RAMP ID='Roof Sheathing_RAMP_Q', T=171.96, F=0.3900/
&RAMP ID='Roof Sheathing_RAMP_Q', T=241.78, F=0.3600/
&RAMP ID='Roof Sheathing_RAMP_Q', T=323.44, F=0.3500/
&RAMP ID='Roof Sheathing_RAMP_Q', T=376.20, F=0.3300/
&RAMP ID='Roof Sheathing_RAMP_Q', T=442.49, F=0.3300/
&RAMP ID='Roof Sheathing_RAMP_Q', T=522.44, F=0.3200/
&RAMP ID='Roof Sheathing_RAMP_Q', T=602.31, F=0.3300/
&RAMP ID='Roof Sheathing_RAMP_Q', T=677.15, F=0.3200/
&RAMP ID='Roof Sheathing_RAMP_Q', T=784.29, F=0.3100/
&RAMP ID='Roof Sheathing_RAMP_Q', T=940.64, F=0.3300/
&RAMP ID='Roof Sheathing_RAMP_Q', T=1.0494200E3, F=0.3400/
&SURF ID='Wood Joists02',
      RGB=168,96,0,
      HRRPUA=181.39,
      RAMP_Q='Wood Joists02_RAMP_Q',
      IGNITION_TEMPERATURE=384.00,
      BACKING='EXPOSED',
      MATL_ID(1,1)='DougFir',
      MATL_MASS_FRACTION(1,1)=1.00,
      THICKNESS(1)=0.0380/
&RAMP ID='Wood Joists02_RAMP_Q', T=0.00, F=0.00/
&RAMP ID='Wood Joists02_RAMP_Q', T=0.1000, F=0.00/
&RAMP ID='Wood Joists02_RAMP_Q', T=0.3800, F=0.3600/
&RAMP ID='Wood Joists02_RAMP_Q', T=1.61, F=0.69/
&RAMP ID='Wood Joists02_RAMP_Q', T=2.52, F=0.94/
&RAMP ID='Wood Joists02_RAMP_Q', T=8.19, F=1.00/
&RAMP ID='Wood Joists02_RAMP_Q', T=10.11, F=0.94/
&RAMP ID='Wood Joists02_RAMP_Q', T=15.38, F=0.89/
&RAMP ID='Wood Joists02_RAMP_Q', T=20.72, F=0.83/
&RAMP ID='Wood Joists02_RAMP_Q', T=22.83, F=0.72/
&RAMP ID='Wood Joists02_RAMP_Q', T=38.51, F=0.62/
&RAMP ID='Wood Joists02_RAMP_Q', T=60.92, F=0.54/
&RAMP ID='Wood Joists02_RAMP_Q', T=91.81, F=0.40600/
&RAMP ID='Wood Joists02_RAMP_Q', T=127.72, F=0.40/
&RAMP ID='Wood Joists02_RAMP_Q', T=171.96, F=0.3900/
&RAMP ID='Wood Joists02_RAMP_Q', T=241.78, F=0.3600/
&RAMP ID='Wood Joists02_RAMP_Q', T=323.44, F=0.3500/
&RAMP ID='Wood Joists02_RAMP_Q', T=376.20, F=0.3300/
&RAMP ID='Wood Joists02_RAMP_Q', T=442.49, F=0.3300/
&RAMP ID='Wood Joists02_RAMP_Q', T=522.44, F=0.3200/
&RAMP ID='Wood Joists02_RAMP_Q', T=602.31, F=0.3300/
&RAMP ID='Wood Joists02_RAMP_Q', T=677.15, F=0.3200/
&RAMP ID='Wood Joists02_RAMP_Q', T=784.29, F=0.3100/
&RAMP ID='Wood Joists02_RAMP_Q', T=940.64, F=0.3300/
&RAMP ID='Wood Joists02_RAMP_Q', T=1.0494200E3, F=0.3400/
&SURF ID='Overhang Matl',
      RGB=204,102,0/
&SURF ID='Glass',
      RGB=0,255,255,
      TRANSPARENCY=0.60,
      BACKING='EXPOSED',
      MATL_ID(1,1)='Glass',
      MATL_MASS_FRACTION(1,1)=1.00,
      THICKNESS(1)=5.00E-3/
&SURF ID='Roof Sheathing02',
      RGB=153,102,0,
      HRRPUA=181.39,
      RAMP_Q='Roof Sheathing02_RAMP_Q',
      IGNITION_TEMPERATURE=384.00,
      TMP_INNER=100.00,
      BACKING='EXPOSED',
      MATL_ID(1,1)='DougFir',
      MATL_MASS_FRACTION(1,1)=1.00,
      THICKNESS(1)=0.0190/
&RAMP ID='Roof Sheathing02_RAMP_Q', T=0.00, F=0.00/
&RAMP ID='Roof Sheathing02_RAMP_Q', T=0.1000, F=0.00/
```

```
&RAMP ID='Roof Sheathing02_RAMP_Q', T=0.3800, F=0.3600/
&RAMP ID='Roof Sheathing02_RAMP_Q', T=1.61, F=0.69/
&RAMP ID='Roof Sheathing02_RAMP_Q', T=2.52, F=0.94/
&RAMP ID='Roof Sheathing02_RAMP_Q', T=8.19, F=1.00/
&RAMP ID='Roof Sheathing02_RAMP_Q', T=10.11, F=0.94/
&RAMP ID='Roof Sheathing02_RAMP_Q', T=15.38, F=0.89/
&RAMP ID='Roof Sheathing02_RAMP_Q', T=20.72, F=0.83/
&RAMP ID='Roof Sheathing02_RAMP_Q', T=22.83, F=0.72/
&RAMP ID='Roof Sheathing02_RAMP_Q', T=38.51, F=0.62/
&RAMP ID='Roof Sheathing02_RAMP_Q', T=60.92, F=0.54/
&RAMP ID='Roof Sheathing02_RAMP_Q', T=91.81, F=0.40600/
&RAMP ID='Roof Sheathing02_RAMP_Q', T=127.72, F=0.40/
&RAMP ID='Roof Sheathing02_RAMP_Q', T=171.96, F=0.3900/
&RAMP ID='Roof Sheathing02_RAMP_Q', T=241.78, F=0.3600/
&RAMP ID='Roof Sheathing02_RAMP_Q', T=323.44, F=0.3500/
&RAMP ID='Roof Sheathing02_RAMP_Q', T=376.20, F=0.3300/
&RAMP ID='Roof Sheathing02_RAMP_Q', T=442.49, F=0.3300/
&RAMP ID='Roof Sheathing02_RAMP_Q', T=522.44, F=0.3200/
&RAMP ID='Roof Sheathing02_RAMP_Q', T=602.31, F=0.3300/
&RAMP ID='Roof Sheathing02_RAMP_Q', T=677.15, F=0.3200/
&RAMP ID='Roof Sheathing02_RAMP_Q', T=784.29, F=0.3100/
&RAMP ID='Roof Sheathing02_RAMP_Q', T=940.64, F=0.3300/
&RAMP ID='Roof Sheathing02_RAMP_Q', T=1.0494200E3, F=0.3400/
&SURF ID='Vent Flow',
      RGB=51,51,204,
      VOLUME_FLUX=0.1750,
      POROUS=.TRUE./
&SURF ID='Cotton',
      RGB=153,153,0,
      MLRPUA=0.0144,
      RAMP_Q='Cotton_RAMP_Q',
      IGNITION_TEMPERATURE=265.00,
      BURN_AWAY=.TRUE.,
      BACKING='EXPOSED',
      MATL_ID(1,1)='Cotton',
      MATL_MASS_FRACTION(1,1)=1.00,
      THICKNESS(1)=0.1000/
&RAMP ID='Cotton_RAMP_Q', T=0.00, F=0.00/
&RAMP ID='Cotton_RAMP_Q', T=5.00, F=0.0107/
&RAMP ID='Cotton_RAMP_Q', T=15.00, F=0.0106/
&RAMP ID='Cotton_RAMP_Q', T=16.00, F=0.1403/
&RAMP ID='Cotton_RAMP_Q', T=21.00, F=1.00/
&RAMP ID='Cotton_RAMP_Q', T=36.00, F=0.99/
&RAMP ID='Cotton_RAMP_Q', T=51.00, F=0.81/
&RAMP ID='Cotton_RAMP_Q', T=74.00, F=0.64/
&RAMP ID='Cotton_RAMP_Q', T=86.00, F=0.61/
&RAMP ID='Cotton_RAMP_Q', T=106.00, F=0.80/
&RAMP ID='Cotton_RAMP_Q', T=123.00, F=0.85/
&RAMP ID='Cotton_RAMP_Q', T=161.00, F=0.75/
&RAMP ID='Cotton_RAMP_Q', T=222.00, F=0.2074/
&RAMP ID='Cotton_RAMP_Q', T=279.00, F=0.0390/
&RAMP ID='Cotton_RAMP_Q', T=307.00, F=0.0385/
&RAMP ID='Cotton_RAMP_Q', T=324.00, F=5.8067250E-3/
&RAMP ID='Cotton_RAMP_Q', T=324.14, F=5.8067250E-3/
&SURF ID='shelf',
      RGB=153,102,0,
      MATL_ID(1,1)='DougFir',
      MATL_MASS_FRACTION(1,1)=1.00,
      THICKNESS(1)=0.0130/
&SURF ID='Brick',
      RGB=102,0,0,
      TEXTURE_MAP='psm_brick.jpg',
      TEXTURE_WIDTH=0.81,
      TEXTURE_HEIGHT=0.81,
      MATL_ID(1,1)='Brick',
      MATL_MASS_FRACTION(1,1)=1.00,
      THICKNESS(1)=0.0900/
&SURF ID='CouchBurner',
      FYI='HRR input determined from Couch Room Burn #19 data from UCF June 2010 LFL experiments',
      COLOR='RED',
      HRRPUA=498.00,
      RAMP_Q='CouchBurner_RAMP_Q'/
&RAMP ID='CouchBurner_RAMP_Q', T=0.00, F=0.00/
&RAMP ID='CouchBurner_RAMP_Q', T=300.00, F=0.00/
&RAMP ID='CouchBurner_RAMP_Q', T=320.00, F=0.0246/
&RAMP ID='CouchBurner_RAMP_Q', T=340.00, F=0.0769/
&RAMP ID='CouchBurner_RAMP_Q', T=360.00, F=0.1384/
&RAMP ID='CouchBurner_RAMP_Q', T=380.00, F=0.1787/
&RAMP ID='CouchBurner_RAMP_Q', T=400.00, F=0.51/
&RAMP ID='CouchBurner_RAMP_Q', T=420.00, F=0.84/
&RAMP ID='CouchBurner_RAMP_Q', T=440.00, F=0.74/
&RAMP ID='CouchBurner_RAMP_Q', T=460.00, F=0.72/
&RAMP ID='CouchBurner_RAMP_Q', T=480.00, F=0.85/
&RAMP ID='CouchBurner_RAMP_Q', T=500.00, F=1.00/
&RAMP ID='CouchBurner_RAMP_Q', T=520.00, F=0.91/
&RAMP ID='CouchBurner_RAMP_Q', T=540.00, F=0.94/
&RAMP ID='CouchBurner_RAMP_Q', T=560.00, F=0.87/
&RAMP ID='CouchBurner_RAMP_Q', T=580.00, F=0.83/
&RAMP ID='CouchBurner_RAMP_Q', T=600.00, F=0.77/
&RAMP ID='CouchBurner_RAMP_Q', T=620.00, F=0.67/
&RAMP ID='CouchBurner_RAMP_Q', T=640.00, F=0.57/
&RAMP ID='CouchBurner_RAMP_Q', T=660.00, F=0.53/
&RAMP ID='CouchBurner_RAMP_Q', T=680.00, F=0.40411/
&RAMP ID='CouchBurner_RAMP_Q', T=700.00, F=0.3767/
&RAMP ID='CouchBurner_RAMP_Q', T=720.00, F=0.3381/
```

```
&RAMP ID='CouchBurner_RAMP_Q', T=740.00, F=0.3075/
&RAMP ID='CouchBurner_RAMP_Q', T=760.00, F=0.2738/
&RAMP ID='CouchBurner_RAMP_Q', T=780.00, F=0.2310/
&RAMP ID='CouchBurner_RAMP_Q', T=800.00, F=0.1974/
&RAMP ID='CouchBurner_RAMP_Q', T=820.00, F=0.1766/
&RAMP ID='CouchBurner_RAMP_Q', T=840.00, F=0.1618/
&RAMP ID='CouchBurner_RAMP_Q', T=860.00, F=0.1457/
&RAMP ID='CouchBurner_RAMP_Q', T=880.00, F=0.1410/
&RAMP ID='CouchBurner_RAMP_Q', T=900.00, F=0.1555/
&RAMP ID='CouchBurner_RAMP_Q', T=920.00, F=0.1453/
&RAMP ID='CouchBurner_RAMP_Q', T=940.00, F=0.1379/
&RAMP ID='CouchBurner_RAMP_Q', T=960.00, F=0.1282/
&RAMP ID='CouchBurner_RAMP_Q', T=980.00, F=0.1328/
&RAMP ID='CouchBurner_RAMP_Q', T=1.00E3, F=0.1193/
&SURF ID='Sofa',
      RGB=51,51,255,
      BACKING='EXPOSED',
      MATL_ID(1,1)='Fiberglass Insulation',
      MATL_MASS_FRACTION(1,1)=1.00,
      THICKNESS(1)=0.1000/
&SURF ID='CouchBurner2',
      FYI='HRR input determined from Couch Room Burn #19 data from UCF June 2010 LFL experiments',
      COLOR='RED',
      HRRPUA=498.00,
      RAMP_Q='CouchBurner2_RAMP_Q'/
&RAMP ID='CouchBurner2_RAMP_Q', T=0.00, F=0.00/
&RAMP ID='CouchBurner2_RAMP_Q', T=640.00, F=0.00/
&RAMP ID='CouchBurner2_RAMP_Q', T=660.00, F=0.0246/
&RAMP ID='CouchBurner2_RAMP_Q', T=680.00, F=0.0769/
&RAMP ID='CouchBurner2_RAMP_Q', T=700.00, F=0.1384/
&RAMP ID='CouchBurner2_RAMP_Q', T=720.00, F=0.1787/
&RAMP ID='CouchBurner2_RAMP_Q', T=740.00, F=0.51/
&RAMP ID='CouchBurner2_RAMP_Q', T=760.00, F=0.84/
&RAMP ID='CouchBurner2_RAMP_Q', T=780.00, F=0.74/
&RAMP ID='CouchBurner2_RAMP_Q', T=800.00, F=0.72/
&RAMP ID='CouchBurner2_RAMP_Q', T=820.00, F=0.85/
&RAMP ID='CouchBurner2_RAMP_Q', T=840.00, F=1.00/
&RAMP ID='CouchBurner2_RAMP_Q', T=860.00, F=0.91/
&RAMP ID='CouchBurner2_RAMP_Q', T=880.00, F=0.94/
&RAMP ID='CouchBurner2_RAMP_Q', T=900.00, F=0.87/
&RAMP ID='CouchBurner2_RAMP_Q', T=920.00, F=0.83/
&RAMP ID='CouchBurner2_RAMP_Q', T=940.00, F=0.77/
&RAMP ID='CouchBurner2_RAMP_Q', T=960.00, F=0.67/
&RAMP ID='CouchBurner2_RAMP_Q', T=980.00, F=0.57/
&RAMP ID='CouchBurner2_RAMP_Q', T=1.00E3, F=0.53/
&RAMP ID='CouchBurner2_RAMP_Q', T=1.02E3, F=0.40411/
&RAMP ID='CouchBurner2_RAMP_Q', T=1.04E3, F=0.3767/
&RAMP ID='CouchBurner2_RAMP_Q', T=1.06E3, F=0.3381/
&RAMP ID='CouchBurner2_RAMP_Q', T=1.08E3, F=0.3075/
&RAMP ID='CouchBurner2_RAMP_Q', T=1.10E3, F=0.2738/
&RAMP ID='CouchBurner2_RAMP_Q', T=1.12E3, F=0.2310/
&RAMP ID='CouchBurner2_RAMP_Q', T=1.14E3, F=0.1974/
&RAMP ID='CouchBurner2_RAMP_Q', T=1.16E3, F=0.1766/
&RAMP ID='CouchBurner2_RAMP_Q', T=1.18E3, F=0.1618/
&RAMP ID='CouchBurner2_RAMP_Q', T=1.20E3, F=0.1457/
&RAMP ID='CouchBurner2_RAMP_Q', T=1.22E3, F=0.1410/
&RAMP ID='CouchBurner2_RAMP_Q', T=1.24E3, F=0.1555/
&RAMP ID='CouchBurner2_RAMP_Q', T=1.26E3, F=0.1453/
&RAMP ID='CouchBurner2_RAMP_Q', T=1.28E3, F=0.1379/
&RAMP ID='CouchBurner2_RAMP_Q', T=1.30E3, F=0.1282/
&RAMP ID='CouchBurner2_RAMP_Q', T=1.32E3, F=0.1328/
&RAMP ID='CouchBurner2_RAMP_Q', T=1.34E3, F=0.1193/
&RAMP ID='CouchBurner2_RAMP_Q', T=1.36E3, F=0.1354/
&RAMP ID='CouchBurner2_RAMP_Q', T=1.38E3, F=0.1159/
&RAMP ID='CouchBurner2_RAMP_Q', T=1.40E3, F=0.1172/
&RAMP ID='CouchBurner2_RAMP_Q', T=1.42E3, F=0.1155/
&RAMP ID='CouchBurner2_RAMP_Q', T=1.44E3, F=0.1168/
&RAMP ID='CouchBurner2_RAMP_Q', T=1.46E3, F=0.1035/
&RAMP ID='CouchBurner2_RAMP_Q', T=1.48E3, F=0.1200/
&RAMP ID='CouchBurner2_RAMP_Q', T=1.50E3, F=0.1191/
&RAMP ID='CouchBurner2_RAMP_Q', T=1.52E3, F=0.1181/
&SURF ID='Wood Floor',
      RGB=153,102,0,
      TEXTURE_MAP='psm_wood2.jpg',
      TEXTURE_WIDTH=0.61,
      TEXTURE_HEIGHT=0.61,
      HRRPUA=181.39,
      RAMP_Q='Wood Floor_RAMP_Q',
      IGNITION_TEMPERATURE=384.00,
      BACKING='EXPOSED',
      MATL_ID(1,1)='DougFir',
      MATL_MASS_FRACTION(1,1)=1.00,
      THICKNESS(1)=0.0190/
&RAMP ID='Wood Floor_RAMP_Q', T=0.00, F=0.00/
&RAMP ID='Wood Floor_RAMP_Q', T=0.1000, F=0.00/
&RAMP ID='Wood Floor_RAMP_Q', T=0.3800, F=0.3600/
&RAMP ID='Wood Floor_RAMP_Q', T=1.61, F=0.69/
&RAMP ID='Wood Floor_RAMP_Q', T=2.52, F=0.94/
&RAMP ID='Wood Floor_RAMP_Q', T=8.19, F=1.00/
&RAMP ID='Wood Floor_RAMP_Q', T=10.11, F=0.94/
&RAMP ID='Wood Floor_RAMP_Q', T=15.38, F=0.89/
&RAMP ID='Wood Floor_RAMP_Q', T=20.72, F=0.83/
&RAMP ID='Wood Floor_RAMP_Q', T=22.83, F=0.72/
&RAMP ID='Wood Floor_RAMP_Q', T=38.51, F=0.62/
&RAMP ID='Wood Floor_RAMP_Q', T=60.92, F=0.54/
```

```
&RAMP ID='Wood Floor_RAMP_Q', T=91.81, F=0.40600/
&RAMP ID='Wood Floor_RAMP_Q', T=127.72, F=0.40/
&RAMP ID='Wood Floor_RAMP_Q', T=171.96, F=0.3900/
&RAMP ID='Wood Floor_RAMP_Q', T=241.78, F=0.3600/
&RAMP ID='Wood Floor_RAMP_Q', T=323.44, F=0.3500/
&RAMP ID='Wood Floor_RAMP_Q', T=376.20, F=0.3300/
&RAMP ID='Wood Floor_RAMP_Q', T=442.49, F=0.3300/
&RAMP ID='Wood Floor_RAMP_Q', T=522.44, F=0.3200/
&RAMP ID='Wood Floor_RAMP_Q', T=602.31, F=0.3300/
&RAMP ID='Wood Floor_RAMP_Q', T=677.15, F=0.3200/
&RAMP ID='Wood Floor_RAMP_Q', T=784.29, F=0.3100/
&RAMP ID='Wood Floor_RAMP_Q', T=940.64, F=0.3300/
&RAMP ID='Wood Floor_RAMP_Q', T=1.0494200E3, F=0.3400/
&RAMP ID='Wood Floor_RAMP_Q', T=1.1650400E3, F=0.3400/
&RAMP ID='Wood Floor_RAMP_Q', T=1.2398600E3, F=0.3300/
&RAMP ID='Wood Floor_RAMP_Q', T=1.3435500E3, F=0.3300/
&RAMP ID='Wood Floor_RAMP_Q', T=1.4115700E3, F=0.3300/
&SURF ID='Tile',
      TEXTURE_MAP='tantile.jpg',
      TEXTURE_WIDTH=0.1000,
      TEXTURE_HEIGHT=0.1000/
&SURF ID='Carpet',
      RGB=153,102,0,
      TEXTURE_MAP='creamcarpet.jpg',
      MATL_ID(1,1)='Carpet',
      MATL_MASS_FRACTION(1,1)=1.00,
      THICKNESS(1)=6.00E-3/
&SURF ID='Concrete',
      COLOR='GRAY 80'/
&SURF ID='Wood Joists',
      RGB=168,96,0,
      HRRPUA=181.39,
      RAMP_Q='Wood Joists_RAMP_Q',
      IGNITION_TEMPERATURE=384.00,
      BACKING='EXPOSED',
      MATL_ID(1,1)='DougFir',
      MATL_MASS_FRACTION(1,1)=1.00,
      THICKNESS(1)=0.0380/
&RAMP ID='Wood Joists_RAMP_Q', T=0.00, F=0.00/
&RAMP ID='Wood Joists_RAMP_Q', T=0.1000, F=0.00/
&RAMP ID='Wood Joists_RAMP_Q', T=0.3800, F=0.3600/
&RAMP ID='Wood Joists_RAMP_Q', T=1.61, F=0.69/
&RAMP ID='Wood Joists_RAMP_Q', T=2.52, F=0.94/
&RAMP ID='Wood Joists_RAMP_Q', T=8.19, F=1.00/
&RAMP ID='Wood Joists_RAMP_Q', T=10.11, F=0.94/
&RAMP ID='Wood Joists_RAMP_Q', T=15.38, F=0.89/
&RAMP ID='Wood Joists_RAMP_Q', T=20.72, F=0.83/
&RAMP ID='Wood Joists_RAMP_Q', T=22.83, F=0.72/
&RAMP ID='Wood Joists_RAMP_Q', T=38.51, F=0.62/
&RAMP ID='Wood Joists_RAMP_Q', T=60.92, F=0.54/
&RAMP ID='Wood Joists_RAMP_Q', T=91.81, F=0.40600/
&RAMP ID='Wood Joists_RAMP_Q', T=127.72, F=0.40/
&RAMP ID='Wood Joists_RAMP_Q', T=171.96, F=0.3900/
&RAMP ID='Wood Joists_RAMP_Q', T=241.78, F=0.3600/
&RAMP ID='Wood Joists_RAMP_Q', T=323.44, F=0.3500/
&RAMP ID='Wood Joists_RAMP_Q', T=376.20, F=0.3300/
&RAMP ID='Wood Joists_RAMP_Q', T=442.49, F=0.3300/
&RAMP ID='Wood Joists_RAMP_Q', T=522.44, F=0.3200/
&RAMP ID='Wood Joists_RAMP_Q', T=602.31, F=0.3300/
&RAMP ID='Wood Joists_RAMP_Q', T=677.15, F=0.3200/
&RAMP ID='Wood Joists_RAMP_Q', T=784.29, F=0.3100/
&RAMP ID='Wood Joists_RAMP_Q', T=940.64, F=0.3300/
&RAMP ID='Wood Joists_RAMP_Q', T=1.0494200E3, F=0.3400/
&RAMP ID='Wood Joists_RAMP_Q', T=1.1650400E3, F=0.3400/
&RAMP ID='Wood Joists_RAMP_Q', T=1.2398600E3, F=0.3300/
&RAMP ID='Wood Joists_RAMP_Q', T=1.3435500E3, F=0.3300/
&RAMP ID='Wood Joists_RAMP_Q', T=1.4115700E3, F=0.3300/
&SURF ID='INTERIORDOOR',
      RGB=146,202,166,
      HRRPUA=181.39,
      RAMP_Q='INTERIORDOOR_RAMP_Q',
      IGNITION_TEMPERATURE=384.00,
      MATL_ID(1,1)='DougFir',
      MATL_MASS_FRACTION(1,1)=1.00,
      THICKNESS(1)=5.00E-3/
&RAMP ID='INTERIORDOOR_RAMP_Q', T=0.00, F=0.00/
&RAMP ID='INTERIORDOOR_RAMP_Q', T=0.1000, F=0.00/
&RAMP ID='INTERIORDOOR_RAMP_Q', T=0.3800, F=0.3600/
&RAMP ID='INTERIORDOOR_RAMP_Q', T=1.61, F=0.69/
&RAMP ID='INTERIORDOOR_RAMP_Q', T=2.52, F=0.94/
&RAMP ID='INTERIORDOOR_RAMP_Q', T=8.19, F=1.00/
&RAMP ID='INTERIORDOOR_RAMP_Q', T=10.11, F=0.94/
&RAMP ID='INTERIORDOOR_RAMP_Q', T=15.38, F=0.89/
&RAMP ID='INTERIORDOOR_RAMP_Q', T=20.72, F=0.83/
&RAMP ID='INTERIORDOOR_RAMP_Q', T=22.83, F=0.72/
&RAMP ID='INTERIORDOOR_RAMP_Q', T=38.51, F=0.62/
&RAMP ID='INTERIORDOOR_RAMP_Q', T=60.92, F=0.54/
&RAMP ID='INTERIORDOOR_RAMP_Q', T=91.81, F=0.40600/
&RAMP ID='INTERIORDOOR_RAMP_Q', T=127.72, F=0.40/
&RAMP ID='INTERIORDOOR_RAMP_Q', T=171.96, F=0.3900/
&RAMP ID='INTERIORDOOR_RAMP_Q', T=241.78, F=0.3600/
&RAMP ID='INTERIORDOOR_RAMP_Q', T=323.44, F=0.3500/
&RAMP ID='INTERIORDOOR_RAMP_Q', T=376.20, F=0.3300/
&RAMP ID='INTERIORDOOR_RAMP_Q', T=442.49, F=0.3300/
&RAMP ID='INTERIORDOOR_RAMP_Q', T=522.44, F=0.3200/
```

```
&RAMP ID='INTERIORDOOR_RAMP_Q', T=602.31, F=0.3300/
&RAMP ID='INTERIORDOOR_RAMP_Q', T=677.15, F=0.3200/
&RAMP ID='INTERIORDOOR_RAMP_Q', T=784.29, F=0.3100/
&RAMP ID='INTERIORDOOR_RAMP_Q', T=940.64, F=0.3300/
&RAMP ID='INTERIORDOOR_RAMP_Q', T=1.0494200E3, F=0.3400/
&RAMP ID='INTERIORDOOR_RAMP_Q', T=1.1650400E3, F=0.3400/
&RAMP ID='INTERIORDOOR_RAMP_Q', T=1.2398600E3, F=0.3300/
&RAMP ID='INTERIORDOOR_RAMP_Q', T=1.3435500E3, F=0.3300/
&RAMP ID='INTERIORDOOR_RAMP_Q', T=1.4115700E3, F=0.3300/
&SURF ID='Wind',
      RGB=51,51,204,
      VEL=-4.47,
      POROUS=.TRUE./
&DEVC ID='HRR', QUANTITY='HRR', XB=-1.60,33.90,0.2000,18.60,2.50,6.00/
&DEVC ID='TwallCouch2', QUANTITY='WALL TEMPERATURE', XYZ=18.30,13.10,0.60, IOR=-2, SETPOINT=265.00/
&DEVC ID='TwallDen', QUANTITY='WALL TEMPERATURE', XYZ=20.20,8.10,3.20, IOR=-1/
&DEVC ID='TIMER', QUANTITY='TIME', XYZ=0.00,0.00,0.00, SETPOINT=800.00, INITIAL_STATE=.TRUE./
&DEVC ID='TIMER2', QUANTITY='TIME', XYZ=0.00,0.00,0.00, SETPOINT=590.00, INITIAL_STATE=.TRUE./
&DEVC ID='TIMER22', QUANTITY='TIME', XYZ=0.00,0.00,0.00, SETPOINT=860.00, INITIAL_STATE=.TRUE./
&DEVC ID='TIMER3', QUANTITY='TIME', XYZ=0.00,0.00,0.00, SETPOINT=855.00, INITIAL_STATE=.TRUE./
&DEVC ID='TIMER4', QUANTITY='TIME', XYZ=0.00,0.00,0.00, SETPOINT=720.00, INITIAL_STATE=.TRUE./
&DEVC ID='TIMER5', QUANTITY='TIME', XYZ=0.00,0.00,0.00, SETPOINT=500.00, INITIAL_STATE=.TRUE./
&OBST XB=20.20,20.50,4.40,5.20,2.40,2.50, RGB=51,51,255, PERMIT_HOLE=.FALSE., SURF_IDS='Attic Gypsum','Ceiling Gypsum','Ceiling
Gypsum', DEVC_ID='TIMER'/ Ceiling_RoofPunch
&OBST XB=20.80,21.10,4.40,5.20,2.40,2.50, RGB=51,51,255, PERMIT_HOLE=.FALSE., SURF_IDS='Attic Gypsum','Ceiling Gypsum','Ceiling
Gypsum', DEVC_ID='TIMER'/ Ceiling_RoofPunch
&OBST XB=20.30,21.50,13.60,17.40,2.40,2.50, SURF_ID='Attic Gypsum'/ Ceiling
&OBST XB=23.70,24.20,6.50,7.00,2.40,2.50, PERMIT_HOLE=.FALSE., SURF_IDS='Attic Burner','INERT','Closet Burner'/ Ceiling_Burner
&OBST XB=21.70,22.40,7.40,8.10,2.40,2.50, COLOR='BLUE', PERMIT_HOLE=.FALSE., SURF_IDS='Attic Gypsum','Gypsum','Gypsum',
DEVC_ID='TIMER'/ Ceiling_E36 Pull
&OBST XB=22.60,24.90,5.60,7.80,2.40,2.50, SURF_IDS='Attic Gypsum','Ceiling Gypsum','Closet Walls'/ Ceiling_Closet
&OBST XB=21.50,22.60,3.40,7.80,2.40,2.50, SURF_IDS='Attic Gypsum','Ceiling Gypsum','Ceiling Gypsum'/ Ceiling_Default
&OBST XB=22.60,24.90,3.40,5.60,2.40,2.50, SURF_IDS='Attic Gypsum','Ceiling Gypsum','Ceiling Gypsum'/ Ceiling_Default
&OBST XB=30.80,33.40,0.40,4.70,2.40,2.50, SURF_ID6='Gypsum','Gypsum','Roof Sheathing','Wood Joists02','Ceiling Gypsum','Attic
Gypsum'/ Ceiling_Default1
&OBST XB=30.80,33.30,9.80,2.00,2.10, SURF_ID6='Gypsum','Gypsum','Wood Joists02','Roof Sheathing','Ceiling Gypsum','Attic
Gypsum'/ Ceiling_Default2
&OBST XB=24.90,30.80,0.40,17.40,2.40,2.50, SURF_ID6='Gypsum','Roof Sheathing','Gypsum','Roof Sheathing','Ceiling Gypsum','Attic
Gypsum'/ Ceiling_Default3
&OBST XB=21.50,24.90,7.80,17.40,2.40,2.50, SURF_ID6='Gypsum','Gypsum','Gypsum','Roof Sheathing','Ceiling Gypsum','Attic Gypsum'/
Ceiling_Default4
&OBST XB=20.20,21.50,2.10,13.60,2.40,2.50, SURF_ID6='Gypsum','Gypsum','Gypsum','Roof Sheathing','Ceiling Gypsum','Attic Gypsum'/
Ceiling_Default5
&OBST XB=15.30,20.20,2.10,7.70,2.40,2.50, SURF_IDS='Attic Gypsum','Gypsum','Ceiling Gypsum'/ Ceiling_Default6
&OBST XB=2.60,15.30,2.10,12.10,2.40,2.50, SURF_IDS='Attic Gypsum','Gypsum','Ceiling Gypsum'/ Ceiling_Default7
&OBST XB=2.60,15.30,12.10,15.00,2.40,2.50, SURF_ID6='Roof Sheathing','Wood Joists02','Gypsum','Roof Sheathing','Ceiling
Gypsum','Attic Gypsum'/ Ceiling_Default8
&OBST XB=0.2000,2.60,2.10,12.10,2.40,2.50, SURF_ID6='Wood Joists02','Gypsum','Gypsum','Roof Sheathing','Ceiling Gypsum','Attic
Gypsum'/ Ceiling_Default9
&OBST XB=15.30,20.30,11.60,12.00,2.50,2.60, SURF_IDS='Attic Gypsum','Gypsum','Ceiling Gypsum'/ Ceiling_LivRm_Vaulted1
&OBST XB=15.30,20.30,11.20,11.60,2.60,2.70, SURF_IDS='Attic Gypsum','Gypsum','Ceiling Gypsum'/ Ceiling_LivRm_Vaulted2
&OBST XB=15.30,20.30,10.80,11.20,2.70,2.80, SURF_IDS='Attic Gypsum','Gypsum','Ceiling Gypsum'/ Ceiling_LivRm_Vaulted3
&OBST XB=15.30,20.30,10.400,10.80,2.80,2.90, SURF_IDS='Attic Gypsum','Gypsum','Ceiling Gypsum'/ Ceiling_LivRm_Vaulted4
&OBST XB=15.30,20.30,10.00,10.400,2.90,3.00, SURF_IDS='Attic Gypsum','Gypsum','Ceiling Gypsum'/ Ceiling_LivRm_Vaulted5
&OBST XB=15.30,20.30,9.60,10.00,3.00,3.10, SURF_IDS='Attic Gypsum','Gypsum','Ceiling Gypsum'/ Ceiling_LivRm_Vaulted6
&OBST XB=15.30,20.30,9.20,9.60,3.10,3.20, SURF_IDS='Attic Gypsum','Gypsum','Ceiling Gypsum'/ Ceiling_LivRm_Vaulted7
&OBST XB=15.30,20.30,8.80,9.20,3.20,3.30, SURF_IDS='Attic Gypsum','Gypsum','Ceiling Gypsum'/ Ceiling_LivRm_Vaulted8
&OBST XB=15.30,20.30,8.40,8.80,3.30,3.40, SURF_IDS='Attic Gypsum','Gypsum','Ceiling Gypsum'/ Ceiling_LivRm_Vaulted9
&OBST XB=15.30,20.30,8.00,8.40,3.40,3.50, SURF_IDS='Attic Gypsum','Gypsum','Ceiling Gypsum'/ Ceiling_LivRm_Vaulted10
&OBST XB=15.30,20.30,7.60,8.00,3.50,3.60, SURF_IDS='Attic Gypsum','Gypsum','Ceiling Gypsum'/ Ceiling_LivRm_Vaulted11
&OBST XB=15.20,20.30,7.60,7.70,2.50,3.60, SURF_IDS='Attic Gypsum','Gypsum','Ceiling Gypsum'/ Ceiling_LivRm_Vaulted12
&OBST XB=15.20,15.30,7.70,11.20,2.50,2.80, SURF_IDS='Attic Gypsum','Gypsum','Ceiling Gypsum'/ Ceiling_LivRm_VaultedB1
&OBST XB=15.20,15.30,7.70,10.80,2.80,2.90, SURF_IDS='Attic Gypsum','Gypsum','Ceiling Gypsum'/ Ceiling_LivRm_VaultedB2
&OBST XB=15.20,15.30,7.70,10.400,3.00, SURF_IDS='Attic Gypsum','Gypsum','Ceiling Gypsum'/ Ceiling_LivRm_VaultedB3
&OBST XB=15.20,15.30,7.70,10.00,3.00,3.10, SURF_IDS='Attic Gypsum','Gypsum','Ceiling Gypsum'/ Ceiling_LivRm_VaultedB4
&OBST XB=15.20,15.30,7.70,9.60,3.10,3.20, SURF_IDS='Attic Gypsum','Gypsum','Ceiling Gypsum'/ Ceiling_LivRm_VaultedB5
&OBST XB=15.20,15.30,7.70,9.20,3.20,3.30, SURF_IDS='Attic Gypsum','Gypsum','Ceiling Gypsum'/ Ceiling_LivRm_VaultedB6
&OBST XB=15.20,15.30,7.70,8.80,3.30,3.40, SURF_IDS='Attic Gypsum','Gypsum','Ceiling Gypsum'/ Ceiling_LivRm_VaultedB7
&OBST XB=15.20,15.30,7.70,8.40,3.40,3.50, SURF_IDS='Attic Gypsum','Gypsum','Ceiling Gypsum'/ Ceiling_LivRm_VaultedB8
&OBST XB=15.20,15.30,7.70,8.00,3.50,3.60, SURF_IDS='Attic Gypsum','Gypsum','Ceiling Gypsum'/ Ceiling_LivRm_VaultedB9
&OBST XB=20.20,20.30,7.70,13.60,2.50,2.70, SURF_IDS='Attic Gypsum','Gypsum','Ceiling Gypsum', DEVC_ID='TIMER2'/
Ceiling_LivRm_VaultedD1
&OBST XB=20.20,20.30,7.70,11.20,2.70,2.80, SURF_IDS='Attic Gypsum','Gypsum','Ceiling Gypsum', DEVC_ID='TIMER2'/
Ceiling_LivRm_VaultedD2
&OBST XB=20.20,20.30,7.70,10.80,2.80,2.90, SURF_IDS='Attic Gypsum','Gypsum','Ceiling Gypsum', DEVC_ID='TIMER2'/
Ceiling_LivRm_VaultedD3
&OBST XB=20.20,20.30,7.70,10.400,2.90,3.00, SURF_IDS='Attic Gypsum','Gypsum','Ceiling Gypsum', DEVC_ID='TIMER2'/
Ceiling_LivRm_VaultedD4
&OBST XB=20.20,20.30,7.70,10.00,3.00,3.10, SURF_IDS='Attic Gypsum','Gypsum','Ceiling Gypsum', DEVC_ID='TIMER2'/
Ceiling_LivRm_VaultedD5
&OBST XB=20.20,20.30,7.70,9.60,3.10,3.20, SURF_IDS='Attic Gypsum','Gypsum','Ceiling Gypsum', DEVC_ID='TIMER2'/
Ceiling_LivRm_VaultedD6
&OBST XB=20.20,20.30,7.70,9.20,3.20,3.30, SURF_IDS='Attic Gypsum','Gypsum','Ceiling Gypsum', DEVC_ID='TIMER2'/
Ceiling_LivRm_VaultedD7
&OBST XB=20.20,20.30,7.70,8.80,3.30,3.40, SURF_IDS='Attic Gypsum','Gypsum','Ceiling Gypsum', DEVC_ID='TIMER2'/
Ceiling_LivRm_VaultedD8
&OBST XB=20.20,20.30,7.70,8.40,3.40,3.50, SURF_IDS='Attic Gypsum','Gypsum','Ceiling Gypsum', DEVC_ID='TIMER2'/
Ceiling_LivRm_VaultedD9
&OBST XB=20.20,20.30,7.70,8.00,3.50,3.60, SURF_IDS='Attic Gypsum','Gypsum','Ceiling Gypsum', DEVC_ID='TIMER2'/
Ceiling_LivRm_VaultedD10
&OBST XB=15.30,20.20,12.00,13.50,2.40,2.50, SURF_IDS='Attic Gypsum','Gypsum','Ceiling Gypsum'/ Ceiling_LivRm_Vaulted1[1]
&OBST XB=30.80,32.30,1.00,1.10,2.50,2.70, SURF_ID='Wood Joists02'/ RafterBathAdd1
&OBST XB=32.30,33.60,1.00,1.10,2.50,2.70, SURF_ID='Wood Joists02'/ RafterBathAdd1
&OBST XB=33.10,33.30,1.00,1.10,2.80,2.90, SURF_ID='Wood Joists02'/ RafterBathAdd1
&OBST XB=33.00,33.20,1.00,1.10,2.90,3.00, SURF_ID='Wood Joists02'/ RafterBathAdd1
```

```
&OBST XB=32.90,33.10,1.00,1.10,3.00,3.10, SURF_ID='Wood Joists02'/ RafterBathAdd1
&OBST XB=32.80,33.00,1.00,1.10,3.10,3.20, SURF_ID='Wood Joists02'/ RafterBathAdd1
&OBST XB=32.70,32.90,1.00,1.10,3.20,3.30, SURF_ID='Wood Joists02'/ RafterBathAdd1
&OBST XB=32.60,32.80,1.00,1.10,3.30,3.40, SURF_ID='Wood Joists02'/ RafterBathAdd1
&OBST XB=32.50,32.70,1.00,1.10,3.40,3.50, SURF_ID='Wood Joists02'/ RafterBathAdd1
&OBST XB=32.40,32.60,1.00,1.10,3.50,3.60, SURF_ID='Wood Joists02'/ RafterBathAdd1
&OBST XB=32.30,32.50,1.00,1.10,3.60,3.70, SURF_ID='Wood Joists02'/ RafterBathAdd1
&OBST XB=32.30,32.40,1.00,1.10,3.70,3.80, SURF_ID='Wood Joists02'/ RafterBathAdd1
&OBST XB=31.10,31.30,1.01,1.11,2.70,2.80, SURF_ID='Wood Joists02'/ RafterBathAdd1
&OBST XB=33.20,33.40,1.00,1.10,2.70,2.80, SURF_ID='Wood Joists02'/ RafterBathAdd1
&OBST XB=31.20,31.40,1.01,1.11,2.80,2.90, SURF_ID='Wood Joists02'/ RafterBathAdd1
&OBST XB=31.30,31.50,1.01,1.11,2.90,3.00, SURF_ID='Wood Joists02'/ RafterBathAdd1
&OBST XB=31.40,31.60,1.01,1.11,3.00,3.10, SURF_ID='Wood Joists02'/ RafterBathAdd1
&OBST XB=31.50,31.70,1.01,1.11,3.10,3.20, SURF_ID='Wood Joists02'/ RafterBathAdd1
&OBST XB=31.60,31.80,1.01,1.11,3.20,3.30, SURF_ID='Wood Joists02'/ RafterBathAdd1
&OBST XB=31.70,31.90,1.01,1.11,3.30,3.40, SURF_ID='Wood Joists02'/ RafterBathAdd1
&OBST XB=31.80,32.00,1.01,1.11,3.40,3.50, SURF_ID='Wood Joists02'/ RafterBathAdd1
&OBST XB=31.90,32.10,1.01,1.11,3.50,3.60, SURF_ID='Wood Joists02'/ RafterBathAdd1
&OBST XB=32.00,32.20,1.01,1.11,3.60,3.70, SURF_ID='Wood Joists02'/ RafterBathAdd1
&OBST XB=32.10,32.20,1.01,1.11,3.70,3.80, SURF_ID='Wood Joists02'/ RafterBathAdd1
&OBST XB=33.20,33.40,1.60,1.70,2.70,2.80, SURF_ID='Wood Joists02'/ RafterBathAdd2
&OBST XB=32.30,33.60,1.60,1.70,2.50,2.70, SURF_ID='Wood Joists02'/ RafterBathAdd2
&OBST XB=33.10,33.30,1.60,1.70,2.80,2.90, SURF_ID='Wood Joists02'/ RafterBathAdd2
&OBST XB=33.00,33.20,1.60,1.70,2.90,3.00, SURF_ID='Wood Joists02'/ RafterBathAdd2
&OBST XB=32.90,33.10,1.60,1.70,3.00,3.10, SURF_ID='Wood Joists02'/ RafterBathAdd2
&OBST XB=32.80,33.00,1.60,1.70,3.10,3.20, SURF_ID='Wood Joists02'/ RafterBathAdd2
&OBST XB=32.70,32.90,1.60,1.70,3.20,3.30, SURF_ID='Wood Joists02'/ RafterBathAdd2
&OBST XB=32.60,32.80,1.60,1.70,3.30,3.40, SURF_ID='Wood Joists02'/ RafterBathAdd2
&OBST XB=32.50,32.70,1.60,1.70,3.40,3.50, SURF_ID='Wood Joists02'/ RafterBathAdd2
&OBST XB=32.40,32.60,1.60,1.70,3.50,3.60, SURF_ID='Wood Joists02'/ RafterBathAdd2
&OBST XB=32.30,32.50,1.60,1.70,3.60,3.70, SURF_ID='Wood Joists02'/ RafterBathAdd2
&OBST XB=32.30,32.40,1.60,1.70,3.70,3.80, SURF_ID='Wood Joists02'/ RafterBathAdd2
&OBST XB=31.10,31.30,1.61,1.71,2.70,2.80, SURF_ID='Wood Joists02'/ RafterBathAdd2
&OBST XB=30.80,32.30,1.60,1.70,2.50,2.70, SURF_ID='Wood Joists02'/ RafterBathAdd2
&OBST XB=31.20,31.40,1.61,1.71,2.80,2.90, SURF_ID='Wood Joists02'/ RafterBathAdd2
&OBST XB=31.30,31.50,1.61,1.71,2.90,3.00, SURF_ID='Wood Joists02'/ RafterBathAdd2
&OBST XB=31.40,31.60,1.61,1.71,3.00,3.10, SURF_ID='Wood Joists02'/ RafterBathAdd2
&OBST XB=31.50,31.70,1.61,1.71,3.10,3.20, SURF_ID='Wood Joists02'/ RafterBathAdd2
&OBST XB=31.60,31.80,1.61,1.71,3.20,3.30, SURF_ID='Wood Joists02'/ RafterBathAdd2
&OBST XB=31.70,31.90,1.61,1.71,3.30,3.40, SURF_ID='Wood Joists02'/ RafterBathAdd2
&OBST XB=31.80,32.00,1.61,1.71,3.40,3.50, SURF_ID='Wood Joists02'/ RafterBathAdd2
&OBST XB=31.90,32.10,1.61,1.71,3.50,3.60, SURF_ID='Wood Joists02'/ RafterBathAdd2
&OBST XB=32.00,32.20,1.61,1.71,3.60,3.70, SURF_ID='Wood Joists02'/ RafterBathAdd2
&OBST XB=32.10,32.20,1.61,1.71,3.70,3.80, SURF_ID='Wood Joists02'/ RafterBathAdd2
&OBST XB=33.20,33.40,2.20,2.30,2.70,2.80, SURF_ID='Wood Joists02'/ RafterBathAdd3
&OBST XB=32.30,33.60,2.20,2.30,2.50,2.70, SURF_ID='Wood Joists02'/ RafterBathAdd3
&OBST XB=33.10,33.30,2.20,2.30,2.80,2.90, SURF_ID='Wood Joists02'/ RafterBathAdd3
&OBST XB=33.00,33.20,2.20,2.30,2.90,3.00, SURF_ID='Wood Joists02'/ RafterBathAdd3
&OBST XB=32.90,33.10,2.20,2.30,3.00,3.10, SURF_ID='Wood Joists02'/ RafterBathAdd3
&OBST XB=32.80,33.00,2.20,2.30,3.10,3.20, SURF_ID='Wood Joists02'/ RafterBathAdd3
&OBST XB=32.70,32.90,2.20,2.30,3.20,3.30, SURF_ID='Wood Joists02'/ RafterBathAdd3
&OBST XB=32.60,32.80,2.20,2.30,3.30,3.40, SURF_ID='Wood Joists02'/ RafterBathAdd3
&OBST XB=32.50,32.70,2.20,2.30,3.40,3.50, SURF_ID='Wood Joists02'/ RafterBathAdd3
&OBST XB=32.40,32.60,2.20,2.30,3.50,3.60, SURF_ID='Wood Joists02'/ RafterBathAdd3
&OBST XB=32.30,32.50,2.20,2.30,3.60,3.70, SURF_ID='Wood Joists02'/ RafterBathAdd3
&OBST XB=32.30,32.40,2.20,2.30,3.70,3.80, SURF_ID='Wood Joists02'/ RafterBathAdd3
&OBST XB=31.10,31.30,2.21,2.31,2.70,2.80, SURF_ID='Wood Joists02'/ RafterBathAdd3
&OBST XB=30.80,32.30,2.20,2.30,2.50,2.70, SURF_ID='Wood Joists02'/ RafterBathAdd3
&OBST XB=31.20,31.40,2.21,2.31,2.80,2.90, SURF_ID='Wood Joists02'/ RafterBathAdd3
&OBST XB=31.30,31.50,2.21,2.31,2.90,3.00, SURF_ID='Wood Joists02'/ RafterBathAdd3
&OBST XB=31.40,31.60,2.21,2.31,3.00,3.10, SURF_ID='Wood Joists02'/ RafterBathAdd3
&OBST XB=31.50,31.70,2.21,2.31,3.10,3.20, SURF_ID='Wood Joists02'/ RafterBathAdd3
&OBST XB=31.60,31.80,2.21,2.31,3.20,3.30, SURF_ID='Wood Joists02'/ RafterBathAdd3
&OBST XB=31.70,31.90,2.21,2.31,3.30,3.40, SURF_ID='Wood Joists02'/ RafterBathAdd3
&OBST XB=31.80,32.00,2.21,2.31,3.40,3.50, SURF_ID='Wood Joists02'/ RafterBathAdd3
&OBST XB=31.90,32.10,2.21,2.31,3.50,3.60, SURF_ID='Wood Joists02'/ RafterBathAdd3
&OBST XB=32.00,32.20,2.21,2.31,3.60,3.70, SURF_ID='Wood Joists02'/ RafterBathAdd3
&OBST XB=32.10,32.20,2.21,2.31,3.70,3.80, SURF_ID='Wood Joists02'/ RafterBathAdd3
&OBST XB=33.20,33.40,2.80,2.90,2.70,2.80, SURF_ID='Wood Joists02'/ RafterBathAdd4
&OBST XB=32.30,33.60,2.80,2.90,2.50,2.70, SURF_ID='Wood Joists02'/ RafterBathAdd4
&OBST XB=33.10,33.30,2.80,2.90,2.80,2.90, SURF_ID='Wood Joists02'/ RafterBathAdd4
&OBST XB=33.00,33.20,2.80,2.90,2.90,3.00, SURF_ID='Wood Joists02'/ RafterBathAdd4
&OBST XB=32.90,33.10,2.80,2.90,3.00,3.10, SURF_ID='Wood Joists02'/ RafterBathAdd4
&OBST XB=32.80,33.00,2.80,2.90,3.10,3.20, SURF_ID='Wood Joists02'/ RafterBathAdd4
&OBST XB=32.70,32.90,2.80,2.90,3.20,3.30, SURF_ID='Wood Joists02'/ RafterBathAdd4
&OBST XB=32.60,32.80,2.80,2.90,3.30,3.40, SURF_ID='Wood Joists02'/ RafterBathAdd4
&OBST XB=32.50,32.70,2.80,2.90,3.40,3.50, SURF_ID='Wood Joists02'/ RafterBathAdd4
&OBST XB=32.40,32.60,2.80,2.90,3.50,3.60, SURF_ID='Wood Joists02'/ RafterBathAdd4
&OBST XB=32.30,32.50,2.80,2.90,3.60,3.70, SURF_ID='Wood Joists02'/ RafterBathAdd4
&OBST XB=32.30,32.40,2.80,2.90,3.70,3.80, SURF_ID='Wood Joists02'/ RafterBathAdd4
&OBST XB=31.10,31.30,2.81,2.91,2.70,2.80, SURF_ID='Wood Joists02'/ RafterBathAdd4
&OBST XB=30.80,32.30,2.80,2.90,2.50,2.70, SURF_ID='Wood Joists02'/ RafterBathAdd4
&OBST XB=31.20,31.40,2.81,2.91,2.80,2.90, SURF_ID='Wood Joists02'/ RafterBathAdd4
&OBST XB=31.30,31.50,2.81,2.91,2.90,3.00, SURF_ID='Wood Joists02'/ RafterBathAdd4
&OBST XB=31.40,31.60,2.81,2.91,3.00,3.10, SURF_ID='Wood Joists02'/ RafterBathAdd4
&OBST XB=31.50,31.70,2.81,2.91,3.10,3.20, SURF_ID='Wood Joists02'/ RafterBathAdd4
&OBST XB=31.60,31.80,2.81,2.91,3.20,3.30, SURF_ID='Wood Joists02'/ RafterBathAdd4
&OBST XB=31.70,31.90,2.81,2.91,3.30,3.40, SURF_ID='Wood Joists02'/ RafterBathAdd4
&OBST XB=31.80,32.00,2.81,2.91,3.40,3.50, SURF_ID='Wood Joists02'/ RafterBathAdd4
&OBST XB=31.90,32.10,2.81,2.91,3.50,3.60, SURF_ID='Wood Joists02'/ RafterBathAdd4
&OBST XB=32.00,32.20,2.81,2.91,3.60,3.70, SURF_ID='Wood Joists02'/ RafterBathAdd4
&OBST XB=32.10,32.20,2.81,2.91,3.70,3.80, SURF_ID='Wood Joists02'/ RafterBathAdd4
&OBST XB=33.20,33.40,3.40,3.50,2.70,2.80, SURF_ID='Wood Joists02'/ RafterBathAdd5
&OBST XB=32.30,33.60,3.40,3.50,2.50,2.70, SURF_ID='Wood Joists02'/ RafterBathAdd5
&OBST XB=33.10,33.30,3.40,3.50,2.80,2.90, SURF_ID='Wood Joists02'/ RafterBathAdd5
```

```
&OBST XB=33.00,33.20,3.40,3.50,2.90,3.00, SURF_ID='Wood Joists02'/ RafterBathAdd5
&OBST XB=32.90,33.10,3.40,3.50,3.00,3.10, SURF_ID='Wood Joists02'/ RafterBathAdd5
&OBST XB=32.80,33.00,3.40,3.50,3.10,3.20, SURF_ID='Wood Joists02'/ RafterBathAdd5
&OBST XB=32.70,32.90,3.40,3.50,3.20,3.30, SURF_ID='Wood Joists02'/ RafterBathAdd5
&OBST XB=32.60,32.80,3.40,3.50,3.30,3.40, SURF_ID='Wood Joists02'/ RafterBathAdd5
&OBST XB=32.50,32.70,3.40,3.50,3.40,3.50, SURF_ID='Wood Joists02'/ RafterBathAdd5
&OBST XB=32.40,32.60,3.40,3.50,3.50,3.60, SURF_ID='Wood Joists02'/ RafterBathAdd5
&OBST XB=32.30,32.50,3.40,3.50,3.60,3.70, SURF_ID='Wood Joists02'/ RafterBathAdd5
&OBST XB=32.30,32.40,3.40,3.50,3.70,3.80, SURF_ID='Wood Joists02'/ RafterBathAdd5
&OBST XB=31.10,31.30,3.41,3.51,2.70,2.80, SURF_ID='Wood Joists02'/ RafterBathAdd5
&OBST XB=30.80,32.30,3.40,3.50,2.50,2.70, SURF_ID='Wood Joists02'/ RafterBathAdd5
&OBST XB=31.20,31.40,3.41,3.51,2.80,2.90, SURF_ID='Wood Joists02'/ RafterBathAdd5
&OBST XB=31.30,31.50,3.41,3.51,2.90,3.00, SURF_ID='Wood Joists02'/ RafterBathAdd5
&OBST XB=31.40,31.60,3.41,3.51,3.00,3.10, SURF_ID='Wood Joists02'/ RafterBathAdd5
&OBST XB=31.50,31.70,3.41,3.51,3.10,3.20, SURF_ID='Wood Joists02'/ RafterBathAdd5
&OBST XB=31.60,31.80,3.41,3.51,3.20,3.30, SURF_ID='Wood Joists02'/ RafterBathAdd5
&OBST XB=31.70,31.90,3.41,3.51,3.30,3.40, SURF_ID='Wood Joists02'/ RafterBathAdd5
&OBST XB=31.80,32.00,3.41,3.51,3.40,3.50, SURF_ID='Wood Joists02'/ RafterBathAdd5
&OBST XB=31.90,32.10,3.41,3.51,3.50,3.60, SURF_ID='Wood Joists02'/ RafterBathAdd5
&OBST XB=32.00,32.20,3.41,3.51,3.60,3.70, SURF_ID='Wood Joists02'/ RafterBathAdd5
&OBST XB=32.10,32.20,3.41,3.51,3.70,3.80, SURF_ID='Wood Joists02'/ RafterBathAdd5
&OBST XB=32.30,33.60,4.00,4.10,2.50,2.70, SURF_ID='Wood Joists02'/ RafterBathAdd6
&OBST XB=33.20,33.40,4.00,4.10,2.70,2.80, SURF_ID='Wood Joists02'/ RafterBathAdd6
&OBST XB=33.10,33.30,4.00,4.10,2.80,2.90, SURF_ID='Wood Joists02'/ RafterBathAdd6
&OBST XB=33.00,33.20,4.00,4.10,2.90,3.00, SURF_ID='Wood Joists02'/ RafterBathAdd6
&OBST XB=32.90,33.10,4.00,4.10,3.00,3.10, SURF_ID='Wood Joists02'/ RafterBathAdd6
&OBST XB=32.80,33.00,4.00,4.10,3.10,3.20, SURF_ID='Wood Joists02'/ RafterBathAdd6
&OBST XB=32.70,32.90,4.00,4.10,3.20,3.30, SURF_ID='Wood Joists02'/ RafterBathAdd6
&OBST XB=32.60,32.80,4.00,4.10,3.30,3.40, SURF_ID='Wood Joists02'/ RafterBathAdd6
&OBST XB=32.50,32.70,4.00,4.10,3.40,3.50, SURF_ID='Wood Joists02'/ RafterBathAdd6
&OBST XB=32.40,32.60,4.00,4.10,3.50,3.60, SURF_ID='Wood Joists02'/ RafterBathAdd6
&OBST XB=32.30,32.50,4.00,4.10,3.60,3.70, SURF_ID='Wood Joists02'/ RafterBathAdd6
&OBST XB=32.30,32.40,4.00,4.10,3.70,3.80, SURF_ID='Wood Joists02'/ RafterBathAdd6
&OBST XB=31.10,31.30,4.01,4.11,2.70,2.80, SURF_ID='Wood Joists02'/ RafterBathAdd6
&OBST XB=30.80,32.30,4.00,4.10,2.50,2.70, SURF_ID='Wood Joists02'/ RafterBathAdd6
&OBST XB=31.20,31.40,4.01,4.11,2.80,2.90, SURF_ID='Wood Joists02'/ RafterBathAdd6
&OBST XB=31.30,31.50,4.01,4.11,2.90,3.00, SURF_ID='Wood Joists02'/ RafterBathAdd6
&OBST XB=31.40,31.60,4.01,4.11,3.00,3.10, SURF_ID='Wood Joists02'/ RafterBathAdd6
&OBST XB=31.50,31.70,4.01,4.11,3.10,3.20, SURF_ID='Wood Joists02'/ RafterBathAdd6
&OBST XB=31.60,31.80,4.01,4.11,3.20,3.30, SURF_ID='Wood Joists02'/ RafterBathAdd6
&OBST XB=31.70,31.90,4.01,4.11,3.30,3.40, SURF_ID='Wood Joists02'/ RafterBathAdd6
&OBST XB=31.80,32.00,4.01,4.11,3.40,3.50, SURF_ID='Wood Joists02'/ RafterBathAdd6
&OBST XB=31.90,32.10,4.01,4.11,3.50,3.60, SURF_ID='Wood Joists02'/ RafterBathAdd6
&OBST XB=32.00,32.20,4.01,4.11,3.60,3.70, SURF_ID='Wood Joists02'/ RafterBathAdd6
&OBST XB=32.10,32.20,4.01,4.11,3.70,3.80, SURF_ID='Wood Joists02'/ RafterBathAdd6
&OBST XB=33.20,33.40,4.60,4.70,2.70,2.80, SURF_ID='Wood Joists02'/ RafterBathAdd7
&OBST XB=32.30,33.60,4.60,4.70,2.50,2.70, SURF_ID='Wood Joists02'/ RafterBathAdd7
&OBST XB=33.10,33.30,4.60,4.70,2.80,2.90, SURF_ID='Wood Joists02'/ RafterBathAdd7
&OBST XB=33.00,33.20,4.60,4.70,2.90,3.00, SURF_ID='Wood Joists02'/ RafterBathAdd7
&OBST XB=32.90,33.10,4.60,4.70,3.00,3.10, SURF_ID='Wood Joists02'/ RafterBathAdd7
&OBST XB=32.80,33.00,4.60,4.70,3.10,3.20, SURF_ID='Wood Joists02'/ RafterBathAdd7
&OBST XB=32.70,32.90,4.60,4.70,3.20,3.30, SURF_ID='Wood Joists02'/ RafterBathAdd7
&OBST XB=32.60,32.80,4.60,4.70,3.30,3.40, SURF_ID='Wood Joists02'/ RafterBathAdd7
&OBST XB=32.50,32.70,4.60,4.70,3.40,3.50, SURF_ID='Wood Joists02'/ RafterBathAdd7
&OBST XB=32.40,32.60,4.60,4.70,3.50,3.60, SURF_ID='Wood Joists02'/ RafterBathAdd7
&OBST XB=32.30,32.50,4.60,4.70,3.60,3.70, SURF_ID='Wood Joists02'/ RafterBathAdd7
&OBST XB=32.30,32.40,4.60,4.70,3.70,3.80, SURF_ID='Wood Joists02'/ RafterBathAdd7
&OBST XB=31.10,31.30,4.61,4.71,2.70,2.80, SURF_ID='Wood Joists02'/ RafterBathAdd7
&OBST XB=30.80,32.30,4.60,4.70,2.50,2.70, SURF_ID='Wood Joists02'/ RafterBathAdd7
&OBST XB=31.20,31.40,4.61,4.71,2.80,2.90, SURF_ID='Wood Joists02'/ RafterBathAdd7
&OBST XB=31.30,31.50,4.61,4.71,2.90,3.00, SURF_ID='Wood Joists02'/ RafterBathAdd7
&OBST XB=31.40,31.60,4.61,4.71,3.00,3.10, SURF_ID='Wood Joists02'/ RafterBathAdd7
&OBST XB=31.50,31.70,4.61,4.71,3.10,3.20, SURF_ID='Wood Joists02'/ RafterBathAdd7
&OBST XB=31.60,31.80,4.61,4.71,3.20,3.30, SURF_ID='Wood Joists02'/ RafterBathAdd7
&OBST XB=31.70,31.90,4.61,4.71,3.30,3.40, SURF_ID='Wood Joists02'/ RafterBathAdd7
&OBST XB=31.80,32.00,4.61,4.71,3.40,3.50, SURF_ID='Wood Joists02'/ RafterBathAdd7
&OBST XB=31.90,32.10,4.61,4.71,3.50,3.60, SURF_ID='Wood Joists02'/ RafterBathAdd7
&OBST XB=32.00,32.20,4.61,4.71,3.60,3.70, SURF_ID='Wood Joists02'/ RafterBathAdd7
&OBST XB=32.10,32.20,4.61,4.71,3.70,3.80, SURF_ID='Wood Joists02'/ RafterBathAdd7
&OBST XB=33.10,33.30,5.20,5.30,2.80,2.90, SURF_ID='Wood Joists02'/ RafterBathAdd8
&OBST XB=33.00,33.20,5.20,5.30,2.90,3.00, SURF_ID='Wood Joists02'/ RafterBathAdd8
&OBST XB=32.90,33.10,5.20,5.30,3.00,3.10, SURF_ID='Wood Joists02'/ RafterBathAdd8
&OBST XB=32.80,33.00,5.20,5.30,3.10,3.20, SURF_ID='Wood Joists02'/ RafterBathAdd8
&OBST XB=32.70,32.90,5.20,5.30,3.20,3.30, SURF_ID='Wood Joists02'/ RafterBathAdd8
&OBST XB=32.60,32.80,5.20,5.30,3.30,3.40, SURF_ID='Wood Joists02'/ RafterBathAdd8
&OBST XB=32.50,32.70,5.20,5.30,3.40,3.50, SURF_ID='Wood Joists02'/ RafterBathAdd8
&OBST XB=32.40,32.60,5.20,5.30,3.50,3.60, SURF_ID='Wood Joists02'/ RafterBathAdd8
&OBST XB=32.30,32.50,5.20,5.30,3.60,3.70, SURF_ID='Wood Joists02'/ RafterBathAdd8
&OBST XB=32.30,32.40,5.20,5.30,3.70,3.80, SURF_ID='Wood Joists02'/ RafterBathAdd8
&OBST XB=31.20,31.40,5.21,5.31,2.80,2.90, SURF_ID='Wood Joists02'/ RafterBathAdd8
&OBST XB=31.30,31.50,5.21,5.31,2.90,3.00, SURF_ID='Wood Joists02'/ RafterBathAdd8
&OBST XB=31.40,31.60,5.21,5.31,3.00,3.10, SURF_ID='Wood Joists02'/ RafterBathAdd8
&OBST XB=31.50,31.70,5.21,5.31,3.10,3.20, SURF_ID='Wood Joists02'/ RafterBathAdd8
&OBST XB=31.60,31.80,5.21,5.31,3.20,3.30, SURF_ID='Wood Joists02'/ RafterBathAdd8
&OBST XB=31.70,31.90,5.21,5.31,3.30,3.40, SURF_ID='Wood Joists02'/ RafterBathAdd8
&OBST XB=31.80,32.00,5.21,5.31,3.40,3.50, SURF_ID='Wood Joists02'/ RafterBathAdd8
&OBST XB=31.90,32.10,5.21,5.31,3.50,3.60, SURF_ID='Wood Joists02'/ RafterBathAdd8
&OBST XB=32.00,32.20,5.21,5.31,3.60,3.70, SURF_ID='Wood Joists02'/ RafterBathAdd8
&OBST XB=32.10,32.20,5.21,5.31,3.70,3.80, SURF_ID='Wood Joists02'/ RafterBathAdd8
&OBST XB=8.30,25.30,6.20,6.30,3.90,4.10, SURF_ID='Wood Joists02'/ RafterRidge1
&OBST XB=4.40,7.90,6.20,6.30,3.90,4.10, SURF_ID='Wood Joists02'/ RafterRidge2
&OBST XB=29.90,30.30,0.60,0.70,2.70,2.80, SURF_ID='Wood Joists02'/ RafterD1
&OBST XB=29.70,29.90,0.60,0.70,2.80,2.90, SURF_ID='Wood Joists02'/ RafterD1
&OBST XB=29.90,30.30,1.00,1.10,2.70,2.80, SURF_ID='Wood Joists02'/ RafterD2
&OBST XB=29.50,30.10,1.00,1.10,2.80,2.90, SURF_ID='Wood Joists02'/ RafterD2
```

```
&OBST XB=29.10,29.50,1.00,1.10,2.90,3.00, SURF_ID='Wood Joists02'/ RafterD2
&OBST XB=28.50,28.70,1.60,1.70,3.10,3.20, SURF_ID='Wood Joists02'/ RafterD3
&OBST XB=29.50,30.10,1.60,1.70,2.80,2.90, SURF_ID='Wood Joists02'/ RafterD3
&OBST XB=29.10,29.50,1.60,1.70,2.90,3.00, SURF_ID='Wood Joists02'/ RafterD3
&OBST XB=28.70,29.10,1.60,1.70,3.00,3.10, SURF_ID='Wood Joists02'/ RafterD3
&OBST XB=29.90,30.30,1.60,1.70,2.70,2.80, SURF_ID='Wood Joists02'/ RafterD3
&OBST XB=29.50,30.10,2.20,2.30,2.80,2.90, SURF_ID='Wood Joists02'/ RafterD4
&OBST XB=29.10,29.50,2.20,2.30,2.90,3.00, SURF_ID='Wood Joists02'/ RafterD4
&OBST XB=28.70,29.10,2.20,2.30,3.00,3.10, SURF_ID='Wood Joists02'/ RafterD4
&OBST XB=28.30,28.70,2.20,2.30,3.10,3.20, SURF_ID='Wood Joists02'/ RafterD4
&OBST XB=27.90,28.30,2.20,2.30,3.20,3.30, SURF_ID='Wood Joists02'/ RafterD4
&OBST XB=29.90,30.30,2.20,2.30,2.70,2.80, SURF_ID='Wood Joists02'/ RafterD4
&OBST XB=29.50,30.10,2.80,2.90,2.80,2.90, SURF_ID='Wood Joists02'/ RafterD5
&OBST XB=29.10,29.50,2.80,2.90,2.90,3.00, SURF_ID='Wood Joists02'/ RafterD5
&OBST XB=28.70,29.10,2.80,2.90,3.00,3.10, SURF_ID='Wood Joists02'/ RafterD5
&OBST XB=28.30,28.70,2.80,2.90,3.10,3.20, SURF_ID='Wood Joists02'/ RafterD5
&OBST XB=27.90,28.30,2.80,2.90,3.20,3.30, SURF_ID='Wood Joists02'/ RafterD5
&OBST XB=27.50,27.90,2.80,2.90,3.30,3.40, SURF_ID='Wood Joists02'/ RafterD5
&OBST XB=27.30,27.50,2.80,2.90,3.40,3.50, SURF_ID='Wood Joists02'/ RafterD5
&OBST XB=29.90,30.30,2.80,2.90,2.70,2.80, SURF_ID='Wood Joists02'/ RafterD5
&OBST XB=29.50,30.10,3.40,3.50,2.80,2.90, SURF_ID='Wood Joists02'/ RafterD6
&OBST XB=29.10,29.50,3.40,3.50,2.90,3.00, SURF_ID='Wood Joists02'/ RafterD6
&OBST XB=28.70,29.10,3.40,3.50,3.00,3.10, SURF_ID='Wood Joists02'/ RafterD6
&OBST XB=28.30,28.70,3.40,3.50,3.10,3.20, SURF_ID='Wood Joists02'/ RafterD6
&OBST XB=27.90,28.30,3.40,3.50,3.20,3.30, SURF_ID='Wood Joists02'/ RafterD6
&OBST XB=27.50,27.90,3.40,3.50,3.30,3.40, SURF_ID='Wood Joists02'/ RafterD6
&OBST XB=27.10,27.50,3.40,3.50,3.40,3.50, SURF_ID='Wood Joists02'/ RafterD6
&OBST XB=26.70,27.10,3.40,3.50,3.50,3.60, SURF_ID='Wood Joists02'/ RafterD6
&OBST XB=26.60,26.70,3.40,3.50,3.60,3.70, SURF_ID='Wood Joists02'/ RafterD6
&OBST XB=29.90,30.30,3.40,3.50,2.70,2.80, SURF_ID='Wood Joists02'/ RafterD6
&OBST XB=29.50,30.10,4.00,4.10,2.80,2.90, SURF_ID='Wood Joists02'/ RafterD7
&OBST XB=29.10,29.50,4.00,4.10,2.90,3.00, SURF_ID='Wood Joists02'/ RafterD7
&OBST XB=28.70,29.10,4.00,4.10,3.00,3.10, SURF_ID='Wood Joists02'/ RafterD7
&OBST XB=28.30,28.70,4.00,4.10,3.10,3.20, SURF_ID='Wood Joists02'/ RafterD7
&OBST XB=27.90,28.30,4.00,4.10,3.20,3.30, SURF_ID='Wood Joists02'/ RafterD7
&OBST XB=27.50,27.90,4.00,4.10,3.30,3.40, SURF_ID='Wood Joists02'/ RafterD7
&OBST XB=27.10,27.50,4.00,4.10,3.40,3.50, SURF_ID='Wood Joists02'/ RafterD7
&OBST XB=26.70,27.10,4.00,4.10,3.50,3.60, SURF_ID='Wood Joists02'/ RafterD7
&OBST XB=26.30,26.70,4.00,4.10,3.60,3.70, SURF_ID='Wood Joists02'/ RafterD7
&OBST XB=26.00,26.30,4.00,4.10,3.70,3.80, SURF_ID='Wood Joists02'/ RafterD7
&OBST XB=29.90,30.30,4.00,4.10,2.70,2.80, SURF_ID='Wood Joists02'/ RafterD7
&OBST XB=29.50,30.10,4.60,4.70,2.80,2.90, SURF_ID='Wood Joists02'/ RafterD8
&OBST XB=29.10,29.50,4.60,4.70,2.90,3.00, SURF_ID='Wood Joists02'/ RafterD8
&OBST XB=28.70,29.10,4.60,4.70,3.00,3.10, SURF_ID='Wood Joists02'/ RafterD8
&OBST XB=28.30,28.70,4.60,4.70,3.10,3.20, SURF_ID='Wood Joists02'/ RafterD8
&OBST XB=27.90,28.30,4.60,4.70,3.20,3.30, SURF_ID='Wood Joists02'/ RafterD8
&OBST XB=27.50,27.90,4.60,4.70,3.30,3.40, SURF_ID='Wood Joists02'/ RafterD8
&OBST XB=27.10,27.50,4.60,4.70,3.40,3.50, SURF_ID='Wood Joists02'/ RafterD8
&OBST XB=26.70,27.10,4.60,4.70,3.50,3.60, SURF_ID='Wood Joists02'/ RafterD8
&OBST XB=26.30,26.70,4.60,4.70,3.60,3.70, SURF_ID='Wood Joists02'/ RafterD8
&OBST XB=25.90,26.30,4.60,4.70,3.70,3.80, SURF_ID='Wood Joists02'/ RafterD8
&OBST XB=25.50,25.90,4.60,4.70,3.80,3.90, SURF_ID='Wood Joists02'/ RafterD8
&OBST XB=29.90,30.30,4.60,4.70,2.70,2.80, SURF_ID='Wood Joists02'/ RafterD8
&OBST XB=29.50,30.10,5.20,5.30,2.80,2.90, SURF_ID='Wood Joists02'/ RafterD9
&OBST XB=29.10,29.50,5.20,5.30,2.90,3.00, SURF_ID='Wood Joists02'/ RafterD9
&OBST XB=28.70,29.10,5.20,5.30,3.00,3.10, SURF_ID='Wood Joists02'/ RafterD9
&OBST XB=28.30,28.70,5.20,5.30,3.10,3.20, SURF_ID='Wood Joists02'/ RafterD9
&OBST XB=27.90,28.30,5.20,5.30,3.20,3.30, SURF_ID='Wood Joists02'/ RafterD9
&OBST XB=27.50,27.90,5.20,5.30,3.30,3.40, SURF_ID='Wood Joists02'/ RafterD9
&OBST XB=27.10,27.50,5.20,5.30,3.40,3.50, SURF_ID='Wood Joists02'/ RafterD9
&OBST XB=26.70,27.10,5.20,5.30,3.50,3.60, SURF_ID='Wood Joists02'/ RafterD9
&OBST XB=26.30,26.70,5.20,5.30,3.60,3.70, SURF_ID='Wood Joists02'/ RafterD9
&OBST XB=25.90,26.30,5.20,5.30,3.70,3.80, SURF_ID='Wood Joists02'/ RafterD9
&OBST XB=25.50,25.90,5.20,5.30,3.80,3.90, SURF_ID='Wood Joists02'/ RafterD9
&OBST XB=29.90,30.30,5.20,5.30,2.70,2.80, SURF_ID='Wood Joists02'/ RafterD9
&OBST XB=29.50,30.10,5.80,5.90,2.80,2.90, SURF_ID='Wood Joists02'/ RafterD10
&OBST XB=29.10,29.50,5.80,5.90,2.90,3.00, SURF_ID='Wood Joists02'/ RafterD10
&OBST XB=28.70,29.10,5.80,5.90,3.00,3.10, SURF_ID='Wood Joists02'/ RafterD10
&OBST XB=28.30,28.70,5.80,5.90,3.10,3.20, SURF_ID='Wood Joists02'/ RafterD10
&OBST XB=27.90,28.30,5.80,5.90,3.20,3.30, SURF_ID='Wood Joists02'/ RafterD10
&OBST XB=27.50,27.90,5.80,5.90,3.30,3.40, SURF_ID='Wood Joists02'/ RafterD10
&OBST XB=27.10,27.50,5.80,5.90,3.40,3.50, SURF_ID='Wood Joists02'/ RafterD10
&OBST XB=26.70,27.10,5.80,5.90,3.50,3.60, SURF_ID='Wood Joists02'/ RafterD10
&OBST XB=26.30,26.70,5.80,5.90,3.60,3.70, SURF_ID='Wood Joists02'/ RafterD10
&OBST XB=25.90,26.30,5.80,5.90,3.70,3.80, SURF_ID='Wood Joists02'/ RafterD10
&OBST XB=25.50,25.90,5.80,5.90,3.80,3.90, SURF_ID='Wood Joists02'/ RafterD10
&OBST XB=29.90,30.30,5.80,5.90,2.70,2.80, SURF_ID='Wood Joists02'/ RafterD10
&OBST XB=29.50,30.10,6.40,6.50,2.80,2.90, SURF_ID='Wood Joists02'/ RafterD11
&OBST XB=29.10,29.50,6.40,6.50,2.90,3.00, SURF_ID='Wood Joists02'/ RafterD11
&OBST XB=28.70,29.10,6.40,6.50,3.00,3.10, SURF_ID='Wood Joists02'/ RafterD11
&OBST XB=28.30,28.70,6.40,6.50,3.10,3.20, SURF_ID='Wood Joists02'/ RafterD11
&OBST XB=27.90,28.30,6.40,6.50,3.20,3.30, SURF_ID='Wood Joists02'/ RafterD11
&OBST XB=27.50,27.90,6.40,6.50,3.30,3.40, SURF_ID='Wood Joists02'/ RafterD11
&OBST XB=27.10,27.50,6.40,6.50,3.40,3.50, SURF_ID='Wood Joists02'/ RafterD11
&OBST XB=26.70,27.10,6.40,6.50,3.50,3.60, SURF_ID='Wood Joists02'/ RafterD11
&OBST XB=26.30,26.70,6.40,6.50,3.60,3.70, SURF_ID='Wood Joists02'/ RafterD11
&OBST XB=25.90,26.30,6.40,6.50,3.70,3.80, SURF_ID='Wood Joists02'/ RafterD11
&OBST XB=25.50,25.90,6.40,6.50,3.80,3.90, SURF_ID='Wood Joists02'/ RafterD11
&OBST XB=29.90,30.30,6.40,6.50,2.70,2.80, SURF_ID='Wood Joists02'/ RafterD11
&OBST XB=29.80,30.20,7.00,7.10,2.70,2.80, SURF_ID='Wood Joists02'/ RafterD12
&OBST XB=29.40,30.00,7.00,7.10,2.80,2.90, SURF_ID='Wood Joists02'/ RafterD12
&OBST XB=29.00,29.40,7.00,7.10,2.90,3.00, SURF_ID='Wood Joists02'/ RafterD12
&OBST XB=28.60,29.00,7.00,7.10,3.00,3.10, SURF_ID='Wood Joists02'/ RafterD12
&OBST XB=28.20,28.60,7.00,7.10,3.10,3.20, SURF_ID='Wood Joists02'/ RafterD12
&OBST XB=27.80,28.20,7.00,7.10,3.20,3.30, SURF_ID='Wood Joists02'/ RafterD12
```

```
&OBST XB=27.40,27.80,7.00,7.10,3.30,3.40, SURF_ID='Wood Joists02'/ RafterD12
&OBST XB=27.00,27.40,7.00,7.10,3.40,3.50, SURF_ID='Wood Joists02'/ RafterD12
&OBST XB=26.60,27.00,7.00,7.10,3.50,3.60, SURF_ID='Wood Joists02'/ RafterD12
&OBST XB=26.20,26.60,7.00,7.10,3.60,3.70, SURF_ID='Wood Joists02'/ RafterD12
&OBST XB=25.80,26.20,7.00,7.10,3.70,3.80, SURF_ID='Wood Joists02'/ RafterD12
&OBST XB=25.40,25.80,7.00,7.10,3.80,3.90, SURF_ID='Wood Joists02'/ RafterD12
&OBST XB=29.80,30.20,8.80,8.90,2.70,2.80, SURF_ID='Wood Joists02'/ RafterD15
&OBST XB=29.40,30.00,8.80,8.90,2.80,2.90, SURF_ID='Wood Joists02'/ RafterD15
&OBST XB=29.00,29.40,8.80,8.90,2.90,3.00, SURF_ID='Wood Joists02'/ RafterD15
&OBST XB=28.60,29.00,8.80,8.90,3.00,3.10, SURF_ID='Wood Joists02'/ RafterD15
&OBST XB=28.20,28.60,8.80,8.90,3.10,3.20, SURF_ID='Wood Joists02'/ RafterD15
&OBST XB=27.80,28.20,8.80,8.90,3.20,3.30, SURF_ID='Wood Joists02'/ RafterD15
&OBST XB=29.80,30.20,9.40,9.50,2.70,2.80, SURF_ID='Wood Joists02'/ RafterD16
&OBST XB=28.40,28.60,9.40,9.50,3.10,3.20, SURF_ID='Wood Joists02'/ RafterD16
&OBST XB=29.40,30.00,9.40,9.50,2.80,2.90, SURF_ID='Wood Joists02'/ RafterD16
&OBST XB=29.00,29.40,9.40,9.50,2.90,3.00, SURF_ID='Wood Joists02'/ RafterD16
&OBST XB=28.60,29.00,9.40,9.50,3.00,3.10, SURF_ID='Wood Joists02'/ RafterD16
&OBST XB=29.80,30.20,10.00,10.10,2.70,2.80, SURF_ID='Wood Joists02'/ RafterD17
&OBST XB=29.40,30.00,10.00,10.10,2.80,2.90, SURF_ID='Wood Joists02'/ RafterD17
&OBST XB=29.00,29.40,10.00,10.10,2.90,3.00, SURF_ID='Wood Joists02'/ RafterD17
&OBST XB=29.80,30.20,10.80,10.90,2.70,2.80, SURF_ID='Wood Joists02'/ RafterD18
&OBST XB=29.60,29.80,10.80,10.90,2.80,2.90, SURF_ID='Wood Joists02'/ RafterD18
&OBST XB=-0.40,-0.2000,10.80,10.90,2.80,2.90, SURF_ID='Wood Joists02'/ RafterB16
&OBST XB=-0.80,-0.40,10.80,10.90,2.70,2.80, SURF_ID='Wood Joists02'/ RafterB16
&OBST XB=0.0,0.40,10.00,10.10,2.90,3.00, SURF_ID='Wood Joists02'/ RafterB15
&OBST XB=-0.40,0.0,10.00,10.10,2.80,2.90, SURF_ID='Wood Joists02'/ RafterB15
&OBST XB=-0.80,-0.40,10.00,10.10,2.70,2.80, SURF_ID='Wood Joists02'/ RafterB15
&OBST XB=0.40,0.80,9.40,9.50,3.00,3.10, SURF_ID='Wood Joists02'/ RafterB14
&OBST XB=0.0,0.40,9.40,9.50,2.90,3.00, SURF_ID='Wood Joists02'/ RafterB14
&OBST XB=-0.40,0.0,9.40,9.50,2.80,2.90, SURF_ID='Wood Joists02'/ RafterB14
&OBST XB=0.80,1.00,9.40,9.50,3.10,3.20, SURF_ID='Wood Joists02'/ RafterB14
&OBST XB=-0.80,-0.40,9.40,9.50,2.70,2.80, SURF_ID='Wood Joists02'/ RafterB14
&OBST XB=1.20,1.60,8.80,8.90,3.20,3.30, SURF_ID='Wood Joists02'/ RafterB13
&OBST XB=0.80,1.20,8.80,8.90,3.10,3.20, SURF_ID='Wood Joists02'/ RafterB13
&OBST XB=0.40,0.80,8.80,8.90,3.00,3.10, SURF_ID='Wood Joists02'/ RafterB13
&OBST XB=0.0,0.40,8.80,8.90,2.90,3.00, SURF_ID='Wood Joists02'/ RafterB13
&OBST XB=-0.40,0.0,8.80,8.90,2.80,2.90, SURF_ID='Wood Joists02'/ RafterB13
&OBST XB=-0.80,-0.40,8.80,8.90,2.70,2.80, SURF_ID='Wood Joists02'/ RafterB13
&OBST XB=2.00,2.20,8.20,8.30,3.40,3.50, SURF_ID='Wood Joists02'/ RafterB12
&OBST XB=1.60,2.00,8.20,8.30,3.30,3.40, SURF_ID='Wood Joists02'/ RafterB12
&OBST XB=1.20,1.60,8.20,8.30,3.20,3.30, SURF_ID='Wood Joists02'/ RafterB12
&OBST XB=0.80,1.20,8.20,8.30,3.10,3.20, SURF_ID='Wood Joists02'/ RafterB12
&OBST XB=0.40,0.80,8.20,8.30,3.00,3.10, SURF_ID='Wood Joists02'/ RafterB12
&OBST XB=0.0,0.40,8.20,8.30,2.90,3.00, SURF_ID='Wood Joists02'/ RafterB12
&OBST XB=-0.40,0.0,8.20,8.30,2.80,2.90, SURF_ID='Wood Joists02'/ RafterB12
&OBST XB=-0.80,-0.40,8.20,8.30,2.70,2.80, SURF_ID='Wood Joists02'/ RafterB12
&OBST XB=2.40,2.80,7.60,7.70,3.50,3.60, SURF_ID='Wood Joists02'/ RafterB11
&OBST XB=2.00,2.40,7.60,7.70,3.40,3.50, SURF_ID='Wood Joists02'/ RafterB11
&OBST XB=1.60,2.00,7.60,7.70,3.30,3.40, SURF_ID='Wood Joists02'/ RafterB11
&OBST XB=1.20,1.60,7.60,7.70,3.20,3.30, SURF_ID='Wood Joists02'/ RafterB11
&OBST XB=0.80,1.20,7.60,7.70,3.10,3.20, SURF_ID='Wood Joists02'/ RafterB11
&OBST XB=0.40,0.80,7.60,7.70,3.00,3.10, SURF_ID='Wood Joists02'/ RafterB11
&OBST XB=0.0,0.40,7.60,7.70,2.90,3.00, SURF_ID='Wood Joists02'/ RafterB11
&OBST XB=-0.40,0.0,7.60,7.70,2.80,2.90, SURF_ID='Wood Joists02'/ RafterB11
&OBST XB=-0.80,-0.40,7.60,7.70,2.70,2.80, SURF_ID='Wood Joists02'/ RafterB11
&OBST XB=3.60,3.90,7.00,7.10,3.80,3.90, SURF_ID='Wood Joists02'/ RafterB10
&OBST XB=3.20,3.60,7.00,7.10,3.70,3.80, SURF_ID='Wood Joists02'/ RafterB10
&OBST XB=2.80,3.20,7.00,7.10,3.60,3.70, SURF_ID='Wood Joists02'/ RafterB10
&OBST XB=2.40,2.80,7.00,7.10,3.50,3.60, SURF_ID='Wood Joists02'/ RafterB10
&OBST XB=2.00,2.40,7.00,7.10,3.40,3.50, SURF_ID='Wood Joists02'/ RafterB10
&OBST XB=1.60,2.00,7.00,7.10,3.30,3.40, SURF_ID='Wood Joists02'/ RafterB10
&OBST XB=1.20,1.60,7.00,7.10,3.20,3.30, SURF_ID='Wood Joists02'/ RafterB10
&OBST XB=0.80,1.20,7.00,7.10,3.10,3.20, SURF_ID='Wood Joists02'/ RafterB10
&OBST XB=0.40,0.80,7.00,7.10,3.00,3.10, SURF_ID='Wood Joists02'/ RafterB10
&OBST XB=0.0,0.40,7.00,7.10,2.90,3.00, SURF_ID='Wood Joists02'/ RafterB10
&OBST XB=-0.40,0.0,7.00,7.10,2.80,2.90, SURF_ID='Wood Joists02'/ RafterB10
&OBST XB=-0.80,-0.40,7.00,7.10,2.70,2.80, SURF_ID='Wood Joists02'/ RafterB10
&OBST XB=-0.80,-0.40,6.40,6.50,2.70,2.80, SURF_ID='Wood Joists02'/ RafterB9
&OBST XB=3.60,3.90,6.40,6.50,3.80,3.90, SURF_ID='Wood Joists02'/ RafterB9
&OBST XB=3.20,3.60,6.40,6.50,3.70,3.80, SURF_ID='Wood Joists02'/ RafterB9
&OBST XB=2.80,3.20,6.40,6.50,3.60,3.70, SURF_ID='Wood Joists02'/ RafterB9
&OBST XB=2.40,2.80,6.40,6.50,3.50,3.60, SURF_ID='Wood Joists02'/ RafterB9
&OBST XB=2.00,2.40,6.40,6.50,3.40,3.50, SURF_ID='Wood Joists02'/ RafterB9
&OBST XB=1.60,2.00,6.40,6.50,3.30,3.40, SURF_ID='Wood Joists02'/ RafterB9
&OBST XB=1.20,1.60,6.40,6.50,3.20,3.30, SURF_ID='Wood Joists02'/ RafterB9
&OBST XB=0.80,1.20,6.40,6.50,3.10,3.20, SURF_ID='Wood Joists02'/ RafterB9
&OBST XB=0.40,0.80,6.40,6.50,3.00,3.10, SURF_ID='Wood Joists02'/ RafterB9
&OBST XB=0.0,0.40,6.40,6.50,2.90,3.00, SURF_ID='Wood Joists02'/ RafterB9
&OBST XB=-0.40,0.0,6.40,6.50,2.80,2.90, SURF_ID='Wood Joists02'/ RafterB9
&OBST XB=-0.80,-0.40,5.80,5.90,2.70,2.80, SURF_ID='Wood Joists02'/ RafterB8
&OBST XB=3.60,3.90,5.80,5.90,3.80,3.90, SURF_ID='Wood Joists02'/ RafterB8
&OBST XB=3.20,3.60,5.80,5.90,3.70,3.80, SURF_ID='Wood Joists02'/ RafterB8
&OBST XB=2.80,3.20,5.80,5.90,3.60,3.70, SURF_ID='Wood Joists02'/ RafterB8
&OBST XB=2.40,2.80,5.80,5.90,3.50,3.60, SURF_ID='Wood Joists02'/ RafterB8
&OBST XB=2.00,2.40,5.80,5.90,3.40,3.50, SURF_ID='Wood Joists02'/ RafterB8
&OBST XB=1.60,2.00,5.80,5.90,3.30,3.40, SURF_ID='Wood Joists02'/ RafterB8
&OBST XB=1.20,1.60,5.80,5.90,3.20,3.30, SURF_ID='Wood Joists02'/ RafterB8
&OBST XB=0.80,1.20,5.80,5.90,3.10,3.20, SURF_ID='Wood Joists02'/ RafterB8
&OBST XB=0.40,0.80,5.80,5.90,3.00,3.10, SURF_ID='Wood Joists02'/ RafterB8
&OBST XB=0.0,0.40,5.80,5.90,2.90,3.00, SURF_ID='Wood Joists02'/ RafterB8
&OBST XB=-0.40,0.0,5.80,5.90,2.80,2.90, SURF_ID='Wood Joists02'/ RafterB8
&OBST XB=3.20,3.60,5.20,5.30,3.70,3.80, SURF_ID='Wood Joists02'/ RafterB7
&OBST XB=2.80,3.20,5.20,5.30,3.60,3.70, SURF_ID='Wood Joists02'/ RafterB7
&OBST XB=2.40,2.80,5.20,5.30,3.50,3.60, SURF_ID='Wood Joists02'/ RafterB7
&OBST XB=2.00,2.40,5.20,5.30,3.40,3.50, SURF_ID='Wood Joists02'/ RafterB7
```

```
&OBST XB=1.60,2.00,5.20,5.30,3.30,3.40, SURF_ID='Wood Joists02'/ RafterB7
&OBST XB=1.20,1.60,5.20,5.30,3.20,3.30, SURF_ID='Wood Joists02'/ RafterB7
&OBST XB=0.80,1.20,5.20,5.30,3.10,3.20, SURF_ID='Wood Joists02'/ RafterB7
&OBST XB=0.40,0.80,5.20,5.30,3.00,3.10, SURF_ID='Wood Joists02'/ RafterB7
&OBST XB=0.0,0.40,5.20,5.30,2.90,3.00, SURF_ID='Wood Joists02'/ RafterB7
&OBST XB=-0.40,0.0,5.20,5.30,2.80,2.90, SURF_ID='Wood Joists02'/ RafterB7
&OBST XB=-0.80,-0.40,5.20,5.30,2.70,2.80, SURF_ID='Wood Joists02'/ RafterB7
&OBST XB=2.40,2.80,4.40,4.50,3.50,3.60, SURF_ID='Wood Joists02'/ RafterB6
&OBST XB=2.00,2.40,4.40,4.50,3.40,3.50, SURF_ID='Wood Joists02'/ RafterB6
&OBST XB=1.60,2.00,4.40,4.50,3.30,3.40, SURF_ID='Wood Joists02'/ RafterB6
&OBST XB=1.20,1.60,4.40,4.50,3.20,3.30, SURF_ID='Wood Joists02'/ RafterB6
&OBST XB=0.80,1.20,4.40,4.50,3.10,3.20, SURF_ID='Wood Joists02'/ RafterB6
&OBST XB=0.40,0.80,4.40,4.50,3.00,3.10, SURF_ID='Wood Joists02'/ RafterB6
&OBST XB=0.0,0.40,4.40,4.50,2.90,3.00, SURF_ID='Wood Joists02'/ RafterB6
&OBST XB=-0.40,0.0,4.40,4.50,2.80,2.90, SURF_ID='Wood Joists02'/ RafterB6
&OBST XB=-0.80,-0.40,4.40,4.50,2.70,2.80, SURF_ID='Wood Joists02'/ RafterB6
&OBST XB=2.00,2.20,3.80,3.90,3.40,3.50, SURF_ID='Wood Joists02'/ RafterB5
&OBST XB=1.60,2.00,3.80,3.90,3.30,3.40, SURF_ID='Wood Joists02'/ RafterB5
&OBST XB=1.20,1.60,3.80,3.90,3.20,3.30, SURF_ID='Wood Joists02'/ RafterB5
&OBST XB=0.80,1.20,3.80,3.90,3.10,3.20, SURF_ID='Wood Joists02'/ RafterB5
&OBST XB=0.40,0.80,3.80,3.90,3.00,3.10, SURF_ID='Wood Joists02'/ RafterB5
&OBST XB=0.0,0.40,3.80,3.90,2.90,3.00, SURF_ID='Wood Joists02'/ RafterB5
&OBST XB=-0.40,0.0,3.80,3.90,2.80,2.90, SURF_ID='Wood Joists02'/ RafterB5
&OBST XB=-0.80,-0.40,3.80,3.90,2.70,2.80, SURF_ID='Wood Joists02'/ RafterB5
&OBST XB=1.20,1.60,3.20,3.30,3.20,3.30, SURF_ID='Wood Joists02'/ RafterB4
&OBST XB=0.80,1.20,3.20,3.30,3.10,3.20, SURF_ID='Wood Joists02'/ RafterB4
&OBST XB=0.40,0.80,3.20,3.30,3.00,3.10, SURF_ID='Wood Joists02'/ RafterB4
&OBST XB=0.0,0.40,3.20,3.30,2.90,3.00, SURF_ID='Wood Joists02'/ RafterB4
&OBST XB=-0.40,0.0,3.20,3.30,2.80,2.90, SURF_ID='Wood Joists02'/ RafterB4
&OBST XB=-0.80,-0.40,3.20,3.30,2.70,2.80, SURF_ID='Wood Joists02'/ RafterB4
&OBST XB=0.40,0.80,2.60,2.70,3.00,3.10, SURF_ID='Wood Joists02'/ RafterB3
&OBST XB=0.0,0.40,2.60,2.70,2.90,3.00, SURF_ID='Wood Joists02'/ RafterB3
&OBST XB=-0.40,0.0,2.60,2.70,2.80,2.90, SURF_ID='Wood Joists02'/ RafterB3
&OBST XB=0.80,1.00,2.60,2.70,3.10,3.20, SURF_ID='Wood Joists02'/ RafterB3
&OBST XB=-0.80,-0.40,2.60,2.70,2.70,2.80, SURF_ID='Wood Joists02'/ RafterB3
&OBST XB=0.0,0.40,2.00,2.10,2.90,3.00, SURF_ID='Wood Joists02'/ RafterB2
&OBST XB=-0.40,0.0,2.00,2.10,2.80,2.90, SURF_ID='Wood Joists02'/ RafterB2
&OBST XB=-0.80,-0.40,2.00,2.10,2.70,2.80, SURF_ID='Wood Joists02'/ RafterB2
&OBST XB=-0.80,-0.2000,1.20,1.30,2.70,2.80, SURF_ID='Wood Joists02'/ RafterB1
&OBST XB=30.30,30.400,-0.50,17.40,2.50,2.70, SURF_ID='Wood Joists02'/ Rafter-8
&OBST XB=29.60,29.70,-0.1,0.3,2.70,2.80, SURF_ID='Wood Joists02'/ Rafter-7
&OBST XB=29.60,29.70,0.3,0.70,2.80,2.90, SURF_ID='Wood Joists02'/ Rafter-7
&OBST XB=29.60,29.70,-0.50,17.40,2.50,2.70, SURF_ID='Wood Joists02'/ Rafter-7
&OBST XB=29.60,29.70,11.20,11.60,2.70,2.80, SURF_ID='Wood Joists02'/ Rafter-7
&OBST XB=29.60,29.70,10.80,11.20,2.80,2.90, SURF_ID='Wood Joists02'/ Rafter-7
&OBST XB=29.00,29.10,-0.1,0.3,2.70,2.80, SURF_ID='Wood Joists02'/ Rafter-6
&OBST XB=29.00,29.10,0.3,0.70,2.80,2.90, SURF_ID='Wood Joists02'/ Rafter-6
&OBST XB=29.00,29.10,0.70,1.10,2.90,3.00, SURF_ID='Wood Joists02'/ Rafter-6
&OBST XB=29.00,29.10,-0.50,17.40,2.50,2.70, SURF_ID='Wood Joists02'/ Rafter-6
&OBST XB=29.00,29.10,11.20,11.60,2.70,2.80, SURF_ID='Wood Joists02'/ Rafter-6
&OBST XB=29.00,29.10,10.80,11.20,2.80,2.90, SURF_ID='Wood Joists02'/ Rafter-6
&OBST XB=29.00,29.10,10.400,10.80,2.90,3.00, SURF_ID='Wood Joists02'/ Rafter-6
&OBST XB=29.00,29.10,10.00,10.400,3.00,3.10, SURF_ID='Wood Joists02'/ Rafter-6
&OBST XB=28.40,28.50,-0.1,0.3,2.70,2.80, SURF_ID='Wood Joists02'/ Rafter-5
&OBST XB=28.40,28.50,0.3,0.70,2.80,2.90, SURF_ID='Wood Joists02'/ Rafter-5
&OBST XB=28.40,28.50,0.70,1.10,2.90,3.00, SURF_ID='Wood Joists02'/ Rafter-5
&OBST XB=28.40,28.50,1.10,1.50,3.00,3.10, SURF_ID='Wood Joists02'/ Rafter-5
&OBST XB=28.40,28.50,1.50,1.70,3.10,3.20, SURF_ID='Wood Joists02'/ Rafter-5
&OBST XB=28.40,28.50,-0.50,17.40,2.50,2.70, SURF_ID='Wood Joists02'/ Rafter-5
&OBST XB=28.40,28.50,11.20,11.60,2.70,2.80, SURF_ID='Wood Joists02'/ Rafter-5
&OBST XB=28.40,28.50,10.80,11.20,2.80,2.90, SURF_ID='Wood Joists02'/ Rafter-5
&OBST XB=28.40,28.50,10.400,10.80,2.90,3.00, SURF_ID='Wood Joists02'/ Rafter-5
&OBST XB=28.40,28.50,10.00,10.400,3.00,3.10, SURF_ID='Wood Joists02'/ Rafter-5
&OBST XB=28.40,28.50,9.40,10.00,3.10,3.20, SURF_ID='Wood Joists02'/ Rafter-5
&OBST XB=27.80,27.90,-0.1,0.3,2.70,2.80, SURF_ID='Wood Joists02'/ Rafter-4
&OBST XB=27.80,27.90,0.3,0.70,2.80,2.90, SURF_ID='Wood Joists02'/ Rafter-4
&OBST XB=27.80,27.90,0.70,1.10,2.90,3.00, SURF_ID='Wood Joists02'/ Rafter-4
&OBST XB=27.80,27.90,1.10,1.50,3.00,3.10, SURF_ID='Wood Joists02'/ Rafter-4
&OBST XB=27.80,27.90,1.50,1.90,3.10,3.20, SURF_ID='Wood Joists02'/ Rafter-4
&OBST XB=27.80,27.90,1.90,2.30,3.20,3.30, SURF_ID='Wood Joists02'/ Rafter-4
&OBST XB=27.80,27.90,-0.50,17.40,2.50,2.70, SURF_ID='Wood Joists02'/ Rafter-4
&OBST XB=27.80,27.90,11.20,11.60,2.70,2.80, SURF_ID='Wood Joists02'/ Rafter-4
&OBST XB=27.80,27.90,10.80,11.20,2.80,2.90, SURF_ID='Wood Joists02'/ Rafter-4
&OBST XB=27.80,27.90,10.400,10.80,2.90,3.00, SURF_ID='Wood Joists02'/ Rafter-4
&OBST XB=27.80,27.90,10.00,10.400,3.00,3.10, SURF_ID='Wood Joists02'/ Rafter-4
&OBST XB=27.80,27.90,9.60,10.00,3.10,3.20, SURF_ID='Wood Joists02'/ Rafter-4
&OBST XB=27.80,27.90,9.20,9.60,3.20,3.30, SURF_ID='Wood Joists02'/ Rafter-4
&OBST XB=27.80,27.90,8.80,9.20,3.30,3.40, SURF_ID='Wood Joists02'/ Rafter-4
&OBST XB=27.20,27.30,-0.1,0.3,2.70,2.80, SURF_ID='Wood Joists02'/ Rafter-3
&OBST XB=27.20,27.30,0.3,0.70,2.80,2.90, SURF_ID='Wood Joists02'/ Rafter-3
&OBST XB=27.20,27.30,0.70,1.10,2.90,3.00, SURF_ID='Wood Joists02'/ Rafter-3
&OBST XB=27.20,27.30,1.10,1.50,3.00,3.10, SURF_ID='Wood Joists02'/ Rafter-3
&OBST XB=27.20,27.30,1.50,1.90,3.10,3.20, SURF_ID='Wood Joists02'/ Rafter-3
&OBST XB=27.20,27.30,1.90,2.30,3.20,3.30, SURF_ID='Wood Joists02'/ Rafter-3
&OBST XB=27.20,27.30,2.30,2.70,3.30,3.40, SURF_ID='Wood Joists02'/ Rafter-3
&OBST XB=27.20,27.30,2.70,2.90,3.40,3.50, SURF_ID='Wood Joists02'/ Rafter-3
&OBST XB=27.20,27.30,-0.50,17.40,2.50,2.70, SURF_ID='Wood Joists02'/ Rafter-3
&OBST XB=27.20,27.30,11.20,11.60,2.70,2.80, SURF_ID='Wood Joists02'/ Rafter-3
&OBST XB=27.20,27.30,10.80,11.20,2.80,2.90, SURF_ID='Wood Joists02'/ Rafter-3
&OBST XB=27.20,27.30,10.400,10.80,2.90,3.00, SURF_ID='Wood Joists02'/ Rafter-3
&OBST XB=27.20,27.30,10.00,10.400,3.00,3.10, SURF_ID='Wood Joists02'/ Rafter-3
&OBST XB=27.20,27.30,9.60,10.00,3.10,3.20, SURF_ID='Wood Joists02'/ Rafter-3
&OBST XB=27.20,27.30,9.20,9.60,3.20,3.30, SURF_ID='Wood Joists02'/ Rafter-3
&OBST XB=27.20,27.30,8.80,9.20,3.30,3.40, SURF_ID='Wood Joists02'/ Rafter-3
&OBST XB=27.20,27.30,8.20,8.80,3.40,3.50, SURF_ID='Wood Joists02'/ Rafter-3
```

```
&OBST XB=26.60,26.70,-0.1,0.3,2.70,2.80, SURF_ID='Wood Joists02'/ Rafter-2
&OBST XB=26.60,26.70,0.3,0.70,2.80,2.90, SURF_ID='Wood Joists02'/ Rafter-2
&OBST XB=26.60,26.70,0.70,1.10,2.90,3.00, SURF_ID='Wood Joists02'/ Rafter-2
&OBST XB=26.60,26.70,1.10,1.50,3.00,3.10, SURF_ID='Wood Joists02'/ Rafter-2
&OBST XB=26.60,26.70,1.50,1.90,3.10,3.20, SURF_ID='Wood Joists02'/ Rafter-2
&OBST XB=26.60,26.70,1.90,2.30,3.20,3.30, SURF_ID='Wood Joists02'/ Rafter-2
&OBST XB=26.60,26.70,2.30,2.70,3.30,3.40, SURF_ID='Wood Joists02'/ Rafter-2
&OBST XB=26.60,26.70,2.70,3.10,3.40,3.50, SURF_ID='Wood Joists02'/ Rafter-2
&OBST XB=26.60,26.70,3.10,3.50,3.50,3.60, SURF_ID='Wood Joists02'/ Rafter-2
&OBST XB=26.60,26.70,-0.50,17.40,2.50,2.70, SURF_ID='Wood Joists02'/ Rafter-2
&OBST XB=26.60,26.70,11.20,11.60,2.70,2.80, SURF_ID='Wood Joists02'/ Rafter-2
&OBST XB=26.60,26.70,10.80,11.20,2.80,2.90, SURF_ID='Wood Joists02'/ Rafter-2
&OBST XB=26.60,26.70,10.400,10.80,2.90,3.00, SURF_ID='Wood Joists02'/ Rafter-2
&OBST XB=26.60,26.70,10.00,10.400,3.00,3.10, SURF_ID='Wood Joists02'/ Rafter-2
&OBST XB=26.60,26.70,9.60,10.00,3.10,3.20, SURF_ID='Wood Joists02'/ Rafter-2
&OBST XB=26.60,26.70,9.20,9.60,3.20,3.30, SURF_ID='Wood Joists02'/ Rafter-2
&OBST XB=26.60,26.70,8.80,9.20,3.30,3.40, SURF_ID='Wood Joists02'/ Rafter-2
&OBST XB=26.60,26.70,8.40,8.80,3.40,3.50, SURF_ID='Wood Joists02'/ Rafter-2
&OBST XB=26.60,26.70,8.00,8.40,3.50,3.60, SURF_ID='Wood Joists02'/ Rafter-2
&OBST XB=26.60,26.70,7.60,8.00,3.60,3.70, SURF_ID='Wood Joists02'/ Rafter-2
&OBST XB=26.00,26.10,-0.1,0.3,2.70,2.80, SURF_ID='Wood Joists02'/ Rafter-1
&OBST XB=26.00,26.10,0.3,0.70,2.80,2.90, SURF_ID='Wood Joists02'/ Rafter-1
&OBST XB=26.00,26.10,0.70,1.10,2.90,3.00, SURF_ID='Wood Joists02'/ Rafter-1
&OBST XB=26.00,26.10,1.10,1.50,3.00,3.10, SURF_ID='Wood Joists02'/ Rafter-1
&OBST XB=26.00,26.10,1.50,1.90,3.10,3.20, SURF_ID='Wood Joists02'/ Rafter-1
&OBST XB=26.00,26.10,1.90,2.30,3.20,3.30, SURF_ID='Wood Joists02'/ Rafter-1
&OBST XB=26.00,26.10,2.30,2.70,3.30,3.40, SURF_ID='Wood Joists02'/ Rafter-1
&OBST XB=26.00,26.10,2.70,3.10,3.40,3.50, SURF_ID='Wood Joists02'/ Rafter-1
&OBST XB=26.00,26.10,3.10,3.50,3.50,3.60, SURF_ID='Wood Joists02'/ Rafter-1
&OBST XB=26.00,26.10,3.50,3.90,3.60,3.70, SURF_ID='Wood Joists02'/ Rafter-1
&OBST XB=26.00,26.10,3.90,4.10,3.70,3.80, SURF_ID='Wood Joists02'/ Rafter-1
&OBST XB=26.00,26.10,-0.50,17.40,2.50,2.70, SURF_ID='Wood Joists02'/ Rafter-1
&OBST XB=26.00,26.10,11.20,11.60,2.70,2.80, SURF_ID='Wood Joists02'/ Rafter-1
&OBST XB=26.00,26.10,10.80,11.20,2.80,2.90, SURF_ID='Wood Joists02'/ Rafter-1
&OBST XB=26.00,26.10,10.400,10.80,2.90,3.00, SURF_ID='Wood Joists02'/ Rafter-1
&OBST XB=26.00,26.10,10.00,10.400,3.00,3.10, SURF_ID='Wood Joists02'/ Rafter-1
&OBST XB=26.00,26.10,9.60,10.00,3.10,3.20, SURF_ID='Wood Joists02'/ Rafter-1
&OBST XB=26.00,26.10,9.20,9.60,3.20,3.30, SURF_ID='Wood Joists02'/ Rafter-1
&OBST XB=26.00,26.10,8.80,9.20,3.30,3.40, SURF_ID='Wood Joists02'/ Rafter-1
&OBST XB=26.00,26.10,8.40,8.80,3.40,3.50, SURF_ID='Wood Joists02'/ Rafter-1
&OBST XB=26.00,26.10,8.00,8.40,3.50,3.60, SURF_ID='Wood Joists02'/ Rafter-1
&OBST XB=26.00,26.10,7.60,8.00,3.60,3.70, SURF_ID='Wood Joists02'/ Rafter-1
&OBST XB=26.00,26.10,7.10,7.60,3.70,3.80, SURF_ID='Wood Joists02'/ Rafter-1
&OBST XB=25.40,25.50,-0.1,0.3,2.70,2.80, SURF_ID='Wood Joists02'/ Rafter0
&OBST XB=25.40,25.50,0.3,0.70,2.80,2.90, SURF_ID='Wood Joists02'/ Rafter0
&OBST XB=25.40,25.50,0.70,1.10,2.90,3.00, SURF_ID='Wood Joists02'/ Rafter0
&OBST XB=25.40,25.50,1.10,1.50,3.00,3.10, SURF_ID='Wood Joists02'/ Rafter0
&OBST XB=25.40,25.50,1.50,1.90,3.10,3.20, SURF_ID='Wood Joists02'/ Rafter0
&OBST XB=25.40,25.50,1.90,2.30,3.20,3.30, SURF_ID='Wood Joists02'/ Rafter0
&OBST XB=25.40,25.50,2.30,2.70,3.30,3.40, SURF_ID='Wood Joists02'/ Rafter0
&OBST XB=25.40,25.50,2.70,3.10,3.40,3.50, SURF_ID='Wood Joists02'/ Rafter0
&OBST XB=25.40,25.50,3.10,3.50,3.50,3.60, SURF_ID='Wood Joists02'/ Rafter0
&OBST XB=25.40,25.50,3.50,3.90,3.60,3.70, SURF_ID='Wood Joists02'/ Rafter0
&OBST XB=25.40,25.50,3.90,4.30,3.70,3.80, SURF_ID='Wood Joists02'/ Rafter0
&OBST XB=25.40,25.50,4.30,4.70,3.80,3.90, SURF_ID='Wood Joists02'/ Rafter0
&OBST XB=25.40,25.50,4.70,6.80,3.80,4.00, SURF_ID='Wood Joists02'/ Rafter0
&OBST XB=25.40,25.50,-0.50,17.40,2.50,2.70, SURF_ID='Wood Joists02'/ Rafter0
&OBST XB=25.40,25.50,11.20,11.60,2.70,2.80, SURF_ID='Wood Joists02'/ Rafter0
&OBST XB=25.40,25.50,10.80,11.20,2.80,2.90, SURF_ID='Wood Joists02'/ Rafter0
&OBST XB=25.40,25.50,10.400,10.80,2.90,3.00, SURF_ID='Wood Joists02'/ Rafter0
&OBST XB=25.40,25.50,10.00,10.400,3.00,3.10, SURF_ID='Wood Joists02'/ Rafter0
&OBST XB=25.40,25.50,9.60,10.00,3.10,3.20, SURF_ID='Wood Joists02'/ Rafter0
&OBST XB=25.40,25.50,9.20,9.60,3.20,3.30, SURF_ID='Wood Joists02'/ Rafter0
&OBST XB=25.40,25.50,8.80,9.20,3.30,3.40, SURF_ID='Wood Joists02'/ Rafter0
&OBST XB=25.40,25.50,8.40,8.80,3.40,3.50, SURF_ID='Wood Joists02'/ Rafter0
&OBST XB=25.40,25.50,8.00,8.40,3.50,3.60, SURF_ID='Wood Joists02'/ Rafter0
&OBST XB=25.40,25.50,7.60,8.00,3.60,3.70, SURF_ID='Wood Joists02'/ Rafter0
&OBST XB=25.40,25.50,7.20,7.60,3.70,3.80, SURF_ID='Wood Joists02'/ Rafter0
&OBST XB=25.40,25.50,6.80,7.20,3.80,3.90, SURF_ID='Wood Joists02'/ Rafter0
&OBST XB=24.80,24.90,0.50,17.40,2.50,2.70, SURF_ID='Wood Joists02'/ Rafter1
&OBST XB=24.80,24.90,0.80,1.30,2.70,2.80, SURF_ID='Wood Joists02'/ Rafter1
&OBST XB=24.80,24.90,1.20,1.70,2.80,2.90, SURF_ID='Wood Joists02'/ Rafter1
&OBST XB=24.80,24.90,1.60,2.10,2.90,3.00, SURF_ID='Wood Joists02'/ Rafter1
&OBST XB=24.80,24.90,2.00,2.50,3.00,3.10, SURF_ID='Wood Joists02'/ Rafter1
&OBST XB=24.80,24.90,2.40,2.90,3.10,3.20, SURF_ID='Wood Joists02'/ Rafter1
&OBST XB=24.80,24.90,2.80,3.30,3.20,3.30, SURF_ID='Wood Joists02'/ Rafter1
&OBST XB=24.80,24.90,3.20,3.70,3.30,3.40, SURF_ID='Wood Joists02'/ Rafter1
&OBST XB=24.80,24.90,3.60,4.10,3.40,3.50, SURF_ID='Wood Joists02'/ Rafter1
&OBST XB=24.80,24.90,4.00,4.50,3.50,3.60, SURF_ID='Wood Joists02'/ Rafter1
&OBST XB=24.80,24.90,4.40,4.90,3.60,3.70, SURF_ID='Wood Joists02'/ Rafter1
&OBST XB=24.80,24.90,4.80,5.30,3.70,3.80, SURF_ID='Wood Joists02'/ Rafter1
&OBST XB=24.80,24.90,5.20,5.70,3.80,3.90, SURF_ID='Wood Joists02'/ Rafter1
&OBST XB=24.80,24.90,5.60,6.20,3.90,4.00, SURF_ID='Wood Joists02'/ Rafter1
&OBST XB=24.80,24.90,6.30,6.90,3.90,4.00, SURF_ID='Wood Joists02'/ Rafter1
&OBST XB=24.80,24.90,6.80,7.30,3.80,3.90, SURF_ID='Wood Joists02'/ Rafter1
&OBST XB=24.80,24.90,7.20,7.70,3.70,3.80, SURF_ID='Wood Joists02'/ Rafter1
&OBST XB=24.80,24.90,7.60,8.10,3.60,3.70, SURF_ID='Wood Joists02'/ Rafter1
&OBST XB=24.80,24.90,8.00,8.50,3.50,3.60, SURF_ID='Wood Joists02'/ Rafter1
&OBST XB=24.80,24.90,8.40,8.90,3.40,3.50, SURF_ID='Wood Joists02'/ Rafter1
&OBST XB=24.80,24.90,8.80,9.30,3.30,3.40, SURF_ID='Wood Joists02'/ Rafter1
&OBST XB=24.80,24.90,9.20,9.70,3.20,3.30, SURF_ID='Wood Joists02'/ Rafter1
&OBST XB=24.80,24.90,9.60,10.10,3.10,3.20, SURF_ID='Wood Joists02'/ Rafter1
&OBST XB=24.80,24.90,10.00,10.50,3.00,3.10, SURF_ID='Wood Joists02'/ Rafter1
&OBST XB=24.80,24.90,10.400,10.90,2.90,3.00, SURF_ID='Wood Joists02'/ Rafter1
&OBST XB=24.80,24.90,10.80,11.30,2.80,2.90, SURF_ID='Wood Joists02'/ Rafter1
```

```
&OBST XB=24.80,24.90,11.20,11.70,2.70,2.80, SURF_ID='Wood Joists02'/ Rafter1
&OBST XB=24.20,24.30,0.80,1.30,2.70,2.80, SURF_ID='Wood Joists02'/ Rafter2
&OBST XB=24.20,24.30,1.20,1.70,2.80,2.90, SURF_ID='Wood Joists02'/ Rafter2
&OBST XB=24.20,24.30,1.60,2.10,2.90,3.00, SURF_ID='Wood Joists02'/ Rafter2
&OBST XB=24.20,24.30,2.00,2.50,3.00,3.10, SURF_ID='Wood Joists02'/ Rafter2
&OBST XB=24.20,24.30,2.40,2.90,3.10,3.20, SURF_ID='Wood Joists02'/ Rafter2
&OBST XB=24.20,24.30,2.80,3.30,3.20,3.30, SURF_ID='Wood Joists02'/ Rafter2
&OBST XB=24.20,24.30,3.20,3.70,3.30,3.40, SURF_ID='Wood Joists02'/ Rafter2
&OBST XB=24.20,24.30,3.60,4.10,3.40,3.50, SURF_ID='Wood Joists02'/ Rafter2
&OBST XB=24.20,24.30,4.00,4.50,3.50,3.60, SURF_ID='Wood Joists02'/ Rafter2
&OBST XB=24.20,24.30,4.40,4.90,3.60,3.70, SURF_ID='Wood Joists02'/ Rafter2
&OBST XB=24.20,24.30,4.80,5.30,3.70,3.80, SURF_ID='Wood Joists02'/ Rafter2
&OBST XB=24.20,24.30,5.20,5.70,3.80,3.90, SURF_ID='Wood Joists02'/ Rafter2
&OBST XB=24.20,24.30,5.60,6.20,3.90,4.00, SURF_ID='Wood Joists02'/ Rafter2
&OBST XB=24.20,24.30,6.30,6.90,3.90,4.00, SURF_ID='Wood Joists02'/ Rafter2
&OBST XB=24.20,24.30,6.80,7.30,3.80,3.90, SURF_ID='Wood Joists02'/ Rafter2
&OBST XB=24.20,24.30,7.20,7.70,3.70,3.80, SURF_ID='Wood Joists02'/ Rafter2
&OBST XB=24.20,24.30,7.60,8.10,3.60,3.70, SURF_ID='Wood Joists02'/ Rafter2
&OBST XB=24.20,24.30,8.00,8.50,3.50,3.60, SURF_ID='Wood Joists02'/ Rafter2
&OBST XB=24.20,24.30,8.40,8.90,3.40,3.50, SURF_ID='Wood Joists02'/ Rafter2
&OBST XB=24.20,24.30,8.80,9.30,3.30,3.40, SURF_ID='Wood Joists02'/ Rafter2
&OBST XB=24.20,24.30,9.20,9.70,3.20,3.30, SURF_ID='Wood Joists02'/ Rafter2
&OBST XB=24.20,24.30,9.60,10.10,3.10,3.20, SURF_ID='Wood Joists02'/ Rafter2
&OBST XB=24.20,24.30,10.00,10.50,3.00,3.10, SURF_ID='Wood Joists02'/ Rafter2
&OBST XB=24.20,24.30,10.400,10.90,2.90,3.00, SURF_ID='Wood Joists02'/ Rafter2
&OBST XB=24.20,24.30,10.80,11.30,2.80,2.90, SURF_ID='Wood Joists02'/ Rafter2
&OBST XB=24.20,24.30,11.20,11.70,2.70,2.80, SURF_ID='Wood Joists02'/ Rafter2
&OBST XB=24.20,24.30,0.50,17.40,2.50,2.70, SURF_ID='Wood Joists02'/ Rafter2
&OBST XB=23.60,23.70,0.50,17.40,2.50,2.70, SURF_ID='Wood Joists02'/ Rafter3
&OBST XB=23.60,23.70,0.80,1.30,2.70,2.80, SURF_ID='Wood Joists02'/ Rafter3
&OBST XB=23.60,23.70,1.20,1.70,2.80,2.90, SURF_ID='Wood Joists02'/ Rafter3
&OBST XB=23.60,23.70,1.60,2.10,2.90,3.00, SURF_ID='Wood Joists02'/ Rafter3
&OBST XB=23.60,23.70,2.00,2.50,3.00,3.10, SURF_ID='Wood Joists02'/ Rafter3
&OBST XB=23.60,23.70,2.40,2.90,3.10,3.20, SURF_ID='Wood Joists02'/ Rafter3
&OBST XB=23.60,23.70,2.80,3.30,3.20,3.30, SURF_ID='Wood Joists02'/ Rafter3
&OBST XB=23.60,23.70,3.20,3.70,3.30,3.40, SURF_ID='Wood Joists02'/ Rafter3
&OBST XB=23.60,23.70,3.60,4.10,3.40,3.50, SURF_ID='Wood Joists02'/ Rafter3
&OBST XB=23.60,23.70,4.00,4.50,3.50,3.60, SURF_ID='Wood Joists02'/ Rafter3
&OBST XB=23.60,23.70,4.40,4.90,3.60,3.70, SURF_ID='Wood Joists02'/ Rafter3
&OBST XB=23.60,23.70,4.80,5.30,3.70,3.80, SURF_ID='Wood Joists02'/ Rafter3
&OBST XB=23.60,23.70,5.20,5.70,3.80,3.90, SURF_ID='Wood Joists02'/ Rafter3
&OBST XB=23.60,23.70,5.60,6.20,3.90,4.00, SURF_ID='Wood Joists02'/ Rafter3
&OBST XB=23.60,23.70,6.30,6.90,3.90,4.00, SURF_ID='Wood Joists02'/ Rafter3
&OBST XB=23.60,23.70,6.80,7.30,3.80,3.90, SURF_ID='Wood Joists02'/ Rafter3
&OBST XB=23.60,23.70,7.20,7.70,3.70,3.80, SURF_ID='Wood Joists02'/ Rafter3
&OBST XB=23.60,23.70,7.60,8.10,3.60,3.70, SURF_ID='Wood Joists02'/ Rafter3
&OBST XB=23.60,23.70,8.00,8.50,3.50,3.60, SURF_ID='Wood Joists02'/ Rafter3
&OBST XB=23.60,23.70,8.40,8.90,3.40,3.50, SURF_ID='Wood Joists02'/ Rafter3
&OBST XB=23.60,23.70,8.80,9.30,3.30,3.40, SURF_ID='Wood Joists02'/ Rafter3
&OBST XB=23.60,23.70,9.20,9.70,3.20,3.30, SURF_ID='Wood Joists02'/ Rafter3
&OBST XB=23.60,23.70,9.60,10.10,3.10,3.20, SURF_ID='Wood Joists02'/ Rafter3
&OBST XB=23.60,23.70,10.00,10.50,3.00,3.10, SURF_ID='Wood Joists02'/ Rafter3
&OBST XB=23.60,23.70,10.400,10.90,2.90,3.00, SURF_ID='Wood Joists02'/ Rafter3
&OBST XB=23.60,23.70,10.80,11.30,2.80,2.90, SURF_ID='Wood Joists02'/ Rafter3
&OBST XB=23.60,23.70,11.20,11.70,2.70,2.80, SURF_ID='Wood Joists02'/ Rafter3
&OBST XB=23.00,23.10,0.50,17.40,2.50,2.70, SURF_ID='Wood Joists02'/ Rafter4
&OBST XB=23.00,23.10,0.80,1.30,2.70,2.80, SURF_ID='Wood Joists02'/ Rafter4
&OBST XB=23.00,23.10,1.20,1.70,2.80,2.90, SURF_ID='Wood Joists02'/ Rafter4
&OBST XB=23.00,23.10,1.60,2.10,2.90,3.00, SURF_ID='Wood Joists02'/ Rafter4
&OBST XB=23.00,23.10,2.00,2.50,3.00,3.10, SURF_ID='Wood Joists02'/ Rafter4
&OBST XB=23.00,23.10,2.40,2.90,3.10,3.20, SURF_ID='Wood Joists02'/ Rafter4
&OBST XB=23.00,23.10,2.80,3.30,3.20,3.30, SURF_ID='Wood Joists02'/ Rafter4
&OBST XB=23.00,23.10,3.20,3.70,3.30,3.40, SURF_ID='Wood Joists02'/ Rafter4
&OBST XB=23.00,23.10,3.60,4.10,3.40,3.50, SURF_ID='Wood Joists02'/ Rafter4
&OBST XB=23.00,23.10,4.00,4.50,3.50,3.60, SURF_ID='Wood Joists02'/ Rafter4
&OBST XB=23.00,23.10,4.40,4.90,3.60,3.70, SURF_ID='Wood Joists02'/ Rafter4
&OBST XB=23.00,23.10,4.80,5.30,3.70,3.80, SURF_ID='Wood Joists02'/ Rafter4
&OBST XB=23.00,23.10,5.20,5.70,3.80,3.90, SURF_ID='Wood Joists02'/ Rafter4
&OBST XB=23.00,23.10,5.60,6.20,3.90,4.00, SURF_ID='Wood Joists02'/ Rafter4
&OBST XB=23.00,23.10,6.30,6.90,3.90,4.00, SURF_ID='Wood Joists02'/ Rafter4
&OBST XB=23.00,23.10,6.80,7.30,3.80,3.90, SURF_ID='Wood Joists02'/ Rafter4
&OBST XB=23.00,23.10,7.20,7.70,3.70,3.80, SURF_ID='Wood Joists02'/ Rafter4
&OBST XB=23.00,23.10,7.60,8.10,3.60,3.70, SURF_ID='Wood Joists02'/ Rafter4
&OBST XB=23.00,23.10,8.00,8.50,3.50,3.60, SURF_ID='Wood Joists02'/ Rafter4
&OBST XB=23.00,23.10,8.40,8.90,3.40,3.50, SURF_ID='Wood Joists02'/ Rafter4
&OBST XB=23.00,23.10,8.80,9.30,3.30,3.40, SURF_ID='Wood Joists02'/ Rafter4
&OBST XB=23.00,23.10,9.20,9.70,3.20,3.30, SURF_ID='Wood Joists02'/ Rafter4
&OBST XB=23.00,23.10,9.60,10.10,3.10,3.20, SURF_ID='Wood Joists02'/ Rafter4
&OBST XB=23.00,23.10,10.00,10.50,3.00,3.10, SURF_ID='Wood Joists02'/ Rafter4
&OBST XB=23.00,23.10,10.400,10.90,2.90,3.00, SURF_ID='Wood Joists02'/ Rafter4
&OBST XB=23.00,23.10,10.80,11.30,2.80,2.90, SURF_ID='Wood Joists02'/ Rafter4
&OBST XB=23.00,23.10,11.20,11.70,2.70,2.80, SURF_ID='Wood Joists02'/ Rafter4
&OBST XB=22.40,22.50,0.50,17.40,2.50,2.70, SURF_ID='Wood Joists02'/ Rafter5
&OBST XB=22.40,22.50,0.80,1.30,2.70,2.80, SURF_ID='Wood Joists02'/ Rafter5
&OBST XB=22.40,22.50,1.20,1.70,2.80,2.90, SURF_ID='Wood Joists02'/ Rafter5
&OBST XB=22.40,22.50,1.60,2.10,2.90,3.00, SURF_ID='Wood Joists02'/ Rafter5
&OBST XB=22.40,22.50,2.00,2.50,3.00,3.10, SURF_ID='Wood Joists02'/ Rafter5
&OBST XB=22.40,22.50,2.40,2.90,3.10,3.20, SURF_ID='Wood Joists02'/ Rafter5
&OBST XB=22.40,22.50,2.80,3.30,3.20,3.30, SURF_ID='Wood Joists02'/ Rafter5
&OBST XB=22.40,22.50,3.20,3.70,3.30,3.40, SURF_ID='Wood Joists02'/ Rafter5
&OBST XB=22.40,22.50,3.60,4.10,3.40,3.50, SURF_ID='Wood Joists02'/ Rafter5
&OBST XB=22.40,22.50,4.00,4.50,3.50,3.60, SURF_ID='Wood Joists02'/ Rafter5
&OBST XB=22.40,22.50,4.40,4.90,3.60,3.70, SURF_ID='Wood Joists02'/ Rafter5
&OBST XB=22.40,22.50,4.80,5.30,3.70,3.80, SURF_ID='Wood Joists02'/ Rafter5
&OBST XB=22.40,22.50,5.20,5.70,3.80,3.90, SURF_ID='Wood Joists02'/ Rafter5
```

```
&OBST XB=22.40,22.50,5.60,6.20,3.90,4.00, SURF_ID='Wood Joists02'/ Rafter5
&OBST XB=22.40,22.50,6.30,6.90,3.90,4.00, SURF_ID='Wood Joists02'/ Rafter5
&OBST XB=22.40,22.50,6.80,7.30,3.80,3.90, SURF_ID='Wood Joists02'/ Rafter5
&OBST XB=22.40,22.50,7.20,7.70,3.70,3.80, SURF_ID='Wood Joists02'/ Rafter5
&OBST XB=22.40,22.50,7.60,8.10,3.60,3.70, SURF_ID='Wood Joists02'/ Rafter5
&OBST XB=22.40,22.50,8.00,8.50,3.50,3.60, SURF_ID='Wood Joists02'/ Rafter5
&OBST XB=22.40,22.50,8.40,8.90,3.40,3.50, SURF_ID='Wood Joists02'/ Rafter5
&OBST XB=22.40,22.50,8.80,9.30,3.30,3.40, SURF_ID='Wood Joists02'/ Rafter5
&OBST XB=22.40,22.50,9.20,9.70,3.20,3.30, SURF_ID='Wood Joists02'/ Rafter5
&OBST XB=22.40,22.50,9.60,10.10,3.10,3.20, SURF_ID='Wood Joists02'/ Rafter5
&OBST XB=22.40,22.50,10.00,10.50,3.00,3.10, SURF_ID='Wood Joists02'/ Rafter5
&OBST XB=22.40,22.50,10.400,10.90,2.90,3.00, SURF_ID='Wood Joists02'/ Rafter5
&OBST XB=22.40,22.50,10.80,11.30,2.80,2.90, SURF_ID='Wood Joists02'/ Rafter5
&OBST XB=22.40,22.50,11.20,11.70,2.70,2.80, SURF_ID='Wood Joists02'/ Rafter5
&OBST XB=21.80,21.90,0.50,17.40,2.50,2.70, SURF_ID='Wood Joists02'/ Rafter6
&OBST XB=21.80,21.90,0.80,1.30,2.70,2.80, SURF_ID='Wood Joists02'/ Rafter6
&OBST XB=21.80,21.90,1.20,1.70,2.80,2.90, SURF_ID='Wood Joists02'/ Rafter6
&OBST XB=21.80,21.90,1.60,2.10,2.90,3.00, SURF_ID='Wood Joists02'/ Rafter6
&OBST XB=21.80,21.90,2.00,2.50,3.00,3.10, SURF_ID='Wood Joists02'/ Rafter6
&OBST XB=21.80,21.90,2.40,2.90,3.10,3.20, SURF_ID='Wood Joists02'/ Rafter6
&OBST XB=21.80,21.90,2.80,3.30,3.20,3.30, SURF_ID='Wood Joists02'/ Rafter6
&OBST XB=21.80,21.90,3.20,3.70,3.30,3.40, SURF_ID='Wood Joists02'/ Rafter6
&OBST XB=21.80,21.90,3.60,4.10,3.40,3.50, SURF_ID='Wood Joists02'/ Rafter6
&OBST XB=21.80,21.90,4.00,4.50,3.50,3.60, SURF_ID='Wood Joists02'/ Rafter6
&OBST XB=21.80,21.90,4.40,4.90,3.60,3.70, SURF_ID='Wood Joists02'/ Rafter6
&OBST XB=21.80,21.90,4.80,5.30,3.70,3.80, SURF_ID='Wood Joists02'/ Rafter6
&OBST XB=21.80,21.90,5.20,5.70,3.80,3.90, SURF_ID='Wood Joists02'/ Rafter6
&OBST XB=21.80,21.90,5.60,6.20,3.90,4.00, SURF_ID='Wood Joists02'/ Rafter6
&OBST XB=21.80,21.90,6.30,6.90,3.90,4.00, SURF_ID='Wood Joists02'/ Rafter6
&OBST XB=21.80,21.90,6.80,7.30,3.80,3.90, SURF_ID='Wood Joists02'/ Rafter6
&OBST XB=21.80,21.90,7.20,7.70,3.70,3.80, SURF_ID='Wood Joists02'/ Rafter6
&OBST XB=21.80,21.90,7.60,8.10,3.60,3.70, SURF_ID='Wood Joists02'/ Rafter6
&OBST XB=21.80,21.90,8.00,8.50,3.50,3.60, SURF_ID='Wood Joists02'/ Rafter6
&OBST XB=21.80,21.90,8.40,8.90,3.40,3.50, SURF_ID='Wood Joists02'/ Rafter6
&OBST XB=21.80,21.90,8.80,9.30,3.30,3.40, SURF_ID='Wood Joists02'/ Rafter6
&OBST XB=21.80,21.90,9.20,9.70,3.20,3.30, SURF_ID='Wood Joists02'/ Rafter6
&OBST XB=21.80,21.90,9.60,10.10,3.10,3.20, SURF_ID='Wood Joists02'/ Rafter6
&OBST XB=21.80,21.90,10.00,10.50,3.00,3.10, SURF_ID='Wood Joists02'/ Rafter6
&OBST XB=21.80,21.90,10.400,10.90,2.90,3.00, SURF_ID='Wood Joists02'/ Rafter6
&OBST XB=21.80,21.90,10.80,11.30,2.80,2.90, SURF_ID='Wood Joists02'/ Rafter6
&OBST XB=21.80,21.90,11.20,11.70,2.70,2.80, SURF_ID='Wood Joists02'/ Rafter6
&OBST XB=21.20,21.30,0.50,17.40,2.50,2.70, SURF_ID='Wood Joists02'/ Rafter7
&OBST XB=21.20,21.30,0.80,1.30,2.70,2.80, SURF_ID='Wood Joists02'/ Rafter7
&OBST XB=21.20,21.30,1.20,1.70,2.80,2.90, SURF_ID='Wood Joists02'/ Rafter7
&OBST XB=21.20,21.30,1.60,2.10,2.90,3.00, SURF_ID='Wood Joists02'/ Rafter7
&OBST XB=21.20,21.30,2.00,2.50,3.00,3.10, SURF_ID='Wood Joists02'/ Rafter7
&OBST XB=21.20,21.30,2.40,2.90,3.10,3.20, SURF_ID='Wood Joists02'/ Rafter7
&OBST XB=21.20,21.30,2.80,3.30,3.20,3.30, SURF_ID='Wood Joists02'/ Rafter7
&OBST XB=21.20,21.30,3.20,3.70,3.30,3.40, SURF_ID='Wood Joists02'/ Rafter7
&OBST XB=21.20,21.30,3.60,4.10,3.40,3.50, SURF_ID='Wood Joists02'/ Rafter7
&OBST XB=21.20,21.30,4.00,4.50,3.50,3.60, SURF_ID='Wood Joists02'/ Rafter7
&OBST XB=21.20,21.30,4.40,4.90,3.60,3.70, SURF_ID='Wood Joists02'/ Rafter7
&OBST XB=21.20,21.30,4.80,5.30,3.70,3.80, SURF_ID='Wood Joists02'/ Rafter7
&OBST XB=21.20,21.30,5.20,5.70,3.80,3.90, SURF_ID='Wood Joists02'/ Rafter7
&OBST XB=21.20,21.30,5.60,6.20,3.90,4.00, SURF_ID='Wood Joists02'/ Rafter7
&OBST XB=21.20,21.30,6.30,6.90,3.90,4.00, SURF_ID='Wood Joists02'/ Rafter7
&OBST XB=21.20,21.30,6.80,7.30,3.80,3.90, SURF_ID='Wood Joists02'/ Rafter7
&OBST XB=21.20,21.30,7.20,7.70,3.70,3.80, SURF_ID='Wood Joists02'/ Rafter7
&OBST XB=21.20,21.30,7.60,8.10,3.60,3.70, SURF_ID='Wood Joists02'/ Rafter7
&OBST XB=21.20,21.30,8.00,8.50,3.50,3.60, SURF_ID='Wood Joists02'/ Rafter7
&OBST XB=21.20,21.30,8.40,8.90,3.40,3.50, SURF_ID='Wood Joists02'/ Rafter7
&OBST XB=21.20,21.30,8.80,9.30,3.30,3.40, SURF_ID='Wood Joists02'/ Rafter7
&OBST XB=21.20,21.30,9.20,9.70,3.20,3.30, SURF_ID='Wood Joists02'/ Rafter7
&OBST XB=21.20,21.30,9.60,10.10,3.10,3.20, SURF_ID='Wood Joists02'/ Rafter7
&OBST XB=21.20,21.30,10.00,10.50,3.00,3.10, SURF_ID='Wood Joists02'/ Rafter7
&OBST XB=21.20,21.30,10.400,10.90,2.90,3.00, SURF_ID='Wood Joists02'/ Rafter7
&OBST XB=21.20,21.30,10.80,11.30,2.80,2.90, SURF_ID='Wood Joists02'/ Rafter7
&OBST XB=21.20,21.30,11.20,11.70,2.70,2.80, SURF_ID='Wood Joists02'/ Rafter7
&OBST XB=20.60,20.70,0.50,17.40,2.50,2.70, SURF_ID='Wood Joists02'/ Rafter8
&OBST XB=20.60,20.70,0.80,1.30,2.70,2.80, SURF_ID='Wood Joists02'/ Rafter8
&OBST XB=20.60,20.70,1.20,1.70,2.80,2.90, SURF_ID='Wood Joists02'/ Rafter8
&OBST XB=20.60,20.70,1.60,2.10,2.90,3.00, SURF_ID='Wood Joists02'/ Rafter8
&OBST XB=20.60,20.70,2.00,2.50,3.00,3.10, SURF_ID='Wood Joists02'/ Rafter8
&OBST XB=20.60,20.70,2.40,2.90,3.10,3.20, SURF_ID='Wood Joists02'/ Rafter8
&OBST XB=20.60,20.70,2.80,3.30,3.20,3.30, SURF_ID='Wood Joists02'/ Rafter8
&OBST XB=20.60,20.70,3.20,3.70,3.30,3.40, SURF_ID='Wood Joists02'/ Rafter8
&OBST XB=20.60,20.70,3.60,4.10,3.40,3.50, SURF_ID='Wood Joists02'/ Rafter8
&OBST XB=20.60,20.70,4.00,4.50,3.50,3.60, SURF_ID='Wood Joists02'/ Rafter8
&OBST XB=20.60,20.70,4.40,4.90,3.60,3.70, SURF_ID='Wood Joists02'/ Rafter8
&OBST XB=20.60,20.70,4.80,5.30,3.70,3.80, SURF_ID='Wood Joists02'/ Rafter8
&OBST XB=20.60,20.70,5.20,5.70,3.80,3.90, SURF_ID='Wood Joists02'/ Rafter8
&OBST XB=20.60,20.70,5.60,6.20,3.90,4.00, SURF_ID='Wood Joists02'/ Rafter8
&OBST XB=20.60,20.70,6.30,6.90,3.90,4.00, SURF_ID='Wood Joists02'/ Rafter8
&OBST XB=20.60,20.70,6.80,7.30,3.80,3.90, SURF_ID='Wood Joists02'/ Rafter8
&OBST XB=20.60,20.70,7.20,7.70,3.70,3.80, SURF_ID='Wood Joists02'/ Rafter8
&OBST XB=20.60,20.70,7.60,8.10,3.60,3.70, SURF_ID='Wood Joists02'/ Rafter8
&OBST XB=20.60,20.70,8.00,8.50,3.50,3.60, SURF_ID='Wood Joists02'/ Rafter8
&OBST XB=20.60,20.70,8.40,8.90,3.40,3.50, SURF_ID='Wood Joists02'/ Rafter8
&OBST XB=20.60,20.70,8.80,9.30,3.30,3.40, SURF_ID='Wood Joists02'/ Rafter8
&OBST XB=20.60,20.70,9.20,9.70,3.20,3.30, SURF_ID='Wood Joists02'/ Rafter8
&OBST XB=20.60,20.70,9.60,10.10,3.10,3.20, SURF_ID='Wood Joists02'/ Rafter8
&OBST XB=20.60,20.70,10.00,10.50,3.00,3.10, SURF_ID='Wood Joists02'/ Rafter8
&OBST XB=20.60,20.70,10.400,10.90,2.90,3.00, SURF_ID='Wood Joists02'/ Rafter8
&OBST XB=20.60,20.70,10.80,11.30,2.80,2.90, SURF_ID='Wood Joists02'/ Rafter8
&OBST XB=20.60,20.70,11.20,11.70,2.70,2.80, SURF_ID='Wood Joists02'/ Rafter8
```

```
&OBST XB=20.00,20.10,0.50,7.60,2.50,2.70, SURF_ID='Wood Joists02'/ Rafter9
&OBST XB=20.00,20.10,0.80,1.30,2.70,2.80, SURF_ID='Wood Joists02'/ Rafter9
&OBST XB=20.00,20.10,1.20,1.70,2.80,2.90, SURF_ID='Wood Joists02'/ Rafter9
&OBST XB=20.00,20.10,1.60,2.10,2.90,3.00, SURF_ID='Wood Joists02'/ Rafter9
&OBST XB=20.00,20.10,2.00,2.50,3.00,3.10, SURF_ID='Wood Joists02'/ Rafter9
&OBST XB=20.00,20.10,2.40,2.90,3.10,3.20, SURF_ID='Wood Joists02'/ Rafter9
&OBST XB=20.00,20.10,2.80,3.30,3.20,3.30, SURF_ID='Wood Joists02'/ Rafter9
&OBST XB=20.00,20.10,3.20,3.70,3.30,3.40, SURF_ID='Wood Joists02'/ Rafter9
&OBST XB=20.00,20.10,3.60,4.10,3.40,3.50, SURF_ID='Wood Joists02'/ Rafter9
&OBST XB=20.00,20.10,4.00,4.50,3.50,3.60, SURF_ID='Wood Joists02'/ Rafter9
&OBST XB=20.00,20.10,4.40,4.90,3.60,3.70, SURF_ID='Wood Joists02'/ Rafter9
&OBST XB=20.00,20.10,4.80,5.30,3.70,3.80, SURF_ID='Wood Joists02'/ Rafter9
&OBST XB=20.00,20.10,5.20,5.70,3.80,3.90, SURF_ID='Wood Joists02'/ Rafter9
&OBST XB=20.00,20.10,5.60,6.20,3.90,4.00, SURF_ID='Wood Joists02'/ Rafter9
&OBST XB=20.00,20.10,6.30,6.90,3.90,4.00, SURF_ID='Wood Joists02'/ Rafter9
&OBST XB=20.00,20.10,6.80,7.30,3.80,3.90, SURF_ID='Wood Joists02'/ Rafter9
&OBST XB=20.00,20.10,7.20,7.70,3.70,3.80, SURF_ID='Wood Joists02'/ Rafter9
&OBST XB=20.00,20.10,7.60,8.10,3.60,3.70, SURF_ID='Wood Joists02'/ Rafter9
&OBST XB=20.00,20.10,8.00,8.50,3.50,3.60, SURF_ID='Wood Joists02'/ Rafter9
&OBST XB=20.00,20.10,8.40,8.90,3.40,3.50, SURF_ID='Wood Joists02'/ Rafter9
&OBST XB=20.00,20.10,8.80,9.30,3.30,3.40, SURF_ID='Wood Joists02'/ Rafter9
&OBST XB=20.00,20.10,9.20,9.70,3.20,3.30, SURF_ID='Wood Joists02'/ Rafter9
&OBST XB=20.00,20.10,9.60,10.10,3.10,3.20, SURF_ID='Wood Joists02'/ Rafter9
&OBST XB=20.00,20.10,10.00,10.50,3.00,3.10, SURF_ID='Wood Joists02'/ Rafter9
&OBST XB=20.00,20.10,10.400,10.90,2.90,3.00, SURF_ID='Wood Joists02'/ Rafter9
&OBST XB=20.00,20.10,10.80,11.30,2.80,2.90, SURF_ID='Wood Joists02'/ Rafter9
&OBST XB=20.00,20.10,11.20,12.80,2.70,2.80, SURF_ID='Wood Joists02'/ Rafter9
&OBST XB=19.40,19.50,0.50,7.60,2.50,2.70, SURF_ID='Wood Joists02'/ Rafter10
&OBST XB=19.40,19.50,0.80,1.30,2.70,2.80, SURF_ID='Wood Joists02'/ Rafter10
&OBST XB=19.40,19.50,1.20,1.70,2.80,2.90, SURF_ID='Wood Joists02'/ Rafter10
&OBST XB=19.40,19.50,1.60,2.10,2.90,3.00, SURF_ID='Wood Joists02'/ Rafter10
&OBST XB=19.40,19.50,2.00,2.50,3.00,3.10, SURF_ID='Wood Joists02'/ Rafter10
&OBST XB=19.40,19.50,2.40,2.90,3.10,3.20, SURF_ID='Wood Joists02'/ Rafter10
&OBST XB=19.40,19.50,2.80,3.30,3.20,3.30, SURF_ID='Wood Joists02'/ Rafter10
&OBST XB=19.40,19.50,3.20,3.70,3.30,3.40, SURF_ID='Wood Joists02'/ Rafter10
&OBST XB=19.40,19.50,3.60,4.10,3.40,3.50, SURF_ID='Wood Joists02'/ Rafter10
&OBST XB=19.40,19.50,4.00,4.50,3.50,3.60, SURF_ID='Wood Joists02'/ Rafter10
&OBST XB=19.40,19.50,4.40,4.90,3.60,3.70, SURF_ID='Wood Joists02'/ Rafter10
&OBST XB=19.40,19.50,4.80,5.30,3.70,3.80, SURF_ID='Wood Joists02'/ Rafter10
&OBST XB=19.40,19.50,5.20,5.70,3.80,3.90, SURF_ID='Wood Joists02'/ Rafter10
&OBST XB=19.40,19.50,5.60,6.20,3.90,4.00, SURF_ID='Wood Joists02'/ Rafter10
&OBST XB=19.40,19.50,6.30,6.90,3.90,4.00, SURF_ID='Wood Joists02'/ Rafter10
&OBST XB=19.40,19.50,6.80,7.30,3.80,3.90, SURF_ID='Wood Joists02'/ Rafter10
&OBST XB=19.40,19.50,7.20,7.70,3.70,3.80, SURF_ID='Wood Joists02'/ Rafter10
&OBST XB=19.40,19.50,7.60,8.10,3.60,3.70, SURF_ID='Wood Joists02'/ Rafter10
&OBST XB=19.40,19.50,8.00,8.50,3.50,3.60, SURF_ID='Wood Joists02'/ Rafter10
&OBST XB=19.40,19.50,8.40,8.90,3.40,3.50, SURF_ID='Wood Joists02'/ Rafter10
&OBST XB=19.40,19.50,8.80,9.30,3.30,3.40, SURF_ID='Wood Joists02'/ Rafter10
&OBST XB=19.40,19.50,9.20,9.70,3.20,3.30, SURF_ID='Wood Joists02'/ Rafter10
&OBST XB=19.40,19.50,9.60,10.10,3.10,3.20, SURF_ID='Wood Joists02'/ Rafter10
&OBST XB=19.40,19.50,10.00,10.50,3.00,3.10, SURF_ID='Wood Joists02'/ Rafter10
&OBST XB=19.40,19.50,10.400,10.90,2.90,3.00, SURF_ID='Wood Joists02'/ Rafter10
&OBST XB=19.40,19.50,10.80,11.30,2.80,2.90, SURF_ID='Wood Joists02'/ Rafter10
&OBST XB=19.40,19.50,11.20,12.80,2.70,2.80, SURF_ID='Wood Joists02'/ Rafter10
&OBST XB=18.80,18.90,0.50,7.60,2.50,2.70, SURF_ID='Wood Joists02'/ Rafter11
&OBST XB=18.80,18.90,0.80,1.30,2.70,2.80, SURF_ID='Wood Joists02'/ Rafter11
&OBST XB=18.80,18.90,1.20,1.70,2.80,2.90, SURF_ID='Wood Joists02'/ Rafter11
&OBST XB=18.80,18.90,1.60,2.10,2.90,3.00, SURF_ID='Wood Joists02'/ Rafter11
&OBST XB=18.80,18.90,2.00,2.50,3.00,3.10, SURF_ID='Wood Joists02'/ Rafter11
&OBST XB=18.80,18.90,2.40,2.90,3.10,3.20, SURF_ID='Wood Joists02'/ Rafter11
&OBST XB=18.80,18.90,2.80,3.30,3.20,3.30, SURF_ID='Wood Joists02'/ Rafter11
&OBST XB=18.80,18.90,3.20,3.70,3.30,3.40, SURF_ID='Wood Joists02'/ Rafter11
&OBST XB=18.80,18.90,3.60,4.10,3.40,3.50, SURF_ID='Wood Joists02'/ Rafter11
&OBST XB=18.80,18.90,4.00,4.50,3.50,3.60, SURF_ID='Wood Joists02'/ Rafter11
&OBST XB=18.80,18.90,4.40,4.90,3.60,3.70, SURF_ID='Wood Joists02'/ Rafter11
&OBST XB=18.80,18.90,4.80,5.30,3.70,3.80, SURF_ID='Wood Joists02'/ Rafter11
&OBST XB=18.80,18.90,5.20,5.70,3.80,3.90, SURF_ID='Wood Joists02'/ Rafter11
&OBST XB=18.80,18.90,5.60,6.20,3.90,4.00, SURF_ID='Wood Joists02'/ Rafter11
&OBST XB=18.80,18.90,6.30,6.90,3.90,4.00, SURF_ID='Wood Joists02'/ Rafter11
&OBST XB=18.80,18.90,6.80,7.30,3.80,3.90, SURF_ID='Wood Joists02'/ Rafter11
&OBST XB=18.80,18.90,7.20,7.70,3.70,3.80, SURF_ID='Wood Joists02'/ Rafter11
&OBST XB=18.80,18.90,7.60,8.10,3.60,3.70, SURF_ID='Wood Joists02'/ Rafter11
&OBST XB=18.80,18.90,8.00,8.50,3.50,3.60, SURF_ID='Wood Joists02'/ Rafter11
&OBST XB=18.80,18.90,8.40,8.90,3.40,3.50, SURF_ID='Wood Joists02'/ Rafter11
&OBST XB=18.80,18.90,8.80,9.30,3.30,3.40, SURF_ID='Wood Joists02'/ Rafter11
&OBST XB=18.80,18.90,9.20,9.70,3.20,3.30, SURF_ID='Wood Joists02'/ Rafter11
&OBST XB=18.80,18.90,9.60,10.10,3.10,3.20, SURF_ID='Wood Joists02'/ Rafter11
&OBST XB=18.80,18.90,10.00,10.50,3.00,3.10, SURF_ID='Wood Joists02'/ Rafter11
&OBST XB=18.80,18.90,10.400,10.90,2.90,3.00, SURF_ID='Wood Joists02'/ Rafter11
&OBST XB=18.80,18.90,10.80,11.30,2.80,2.90, SURF_ID='Wood Joists02'/ Rafter11
&OBST XB=18.80,18.90,11.20,12.80,2.70,2.80, SURF_ID='Wood Joists02'/ Rafter11
&OBST XB=18.20,18.30,0.80,1.30,2.70,2.80, SURF_ID='Wood Joists02'/ Rafter12
&OBST XB=18.20,18.30,1.20,1.70,2.80,2.90, SURF_ID='Wood Joists02'/ Rafter12
&OBST XB=18.20,18.30,1.60,2.10,2.90,3.00, SURF_ID='Wood Joists02'/ Rafter12
&OBST XB=18.20,18.30,2.00,2.50,3.00,3.10, SURF_ID='Wood Joists02'/ Rafter12
&OBST XB=18.20,18.30,2.40,2.90,3.10,3.20, SURF_ID='Wood Joists02'/ Rafter12
&OBST XB=18.20,18.30,2.80,3.30,3.20,3.30, SURF_ID='Wood Joists02'/ Rafter12
&OBST XB=18.20,18.30,3.20,3.70,3.30,3.40, SURF_ID='Wood Joists02'/ Rafter12
&OBST XB=18.20,18.30,3.60,4.10,3.40,3.50, SURF_ID='Wood Joists02'/ Rafter12
&OBST XB=18.20,18.30,4.00,4.50,3.50,3.60, SURF_ID='Wood Joists02'/ Rafter12
&OBST XB=18.20,18.30,4.40,4.90,3.60,3.70, SURF_ID='Wood Joists02'/ Rafter12
&OBST XB=18.20,18.30,4.80,5.30,3.70,3.80, SURF_ID='Wood Joists02'/ Rafter12
&OBST XB=18.20,18.30,5.20,5.70,3.80,3.90, SURF_ID='Wood Joists02'/ Rafter12
&OBST XB=18.20,18.30,5.60,6.20,3.90,4.00, SURF_ID='Wood Joists02'/ Rafter12
&OBST XB=18.20,18.30,6.30,6.90,3.90,4.00, SURF_ID='Wood Joists02'/ Rafter12
```

```
&OBST XB=18.20,18.30,6.80,7.30,3.80,3.90, SURF_ID='Wood Joists02'/ Rafter12
&OBST XB=18.20,18.30,7.20,7.70,3.70,3.80, SURF_ID='Wood Joists02'/ Rafter12
&OBST XB=18.20,18.30,7.60,8.10,3.60,3.70, SURF_ID='Wood Joists02'/ Rafter12
&OBST XB=18.20,18.30,8.00,8.50,3.50,3.60, SURF_ID='Wood Joists02'/ Rafter12
&OBST XB=18.20,18.30,8.40,8.90,3.40,3.50, SURF_ID='Wood Joists02'/ Rafter12
&OBST XB=18.20,18.30,8.80,9.30,3.30,3.40, SURF_ID='Wood Joists02'/ Rafter12
&OBST XB=18.20,18.30,9.20,9.70,3.20,3.30, SURF_ID='Wood Joists02'/ Rafter12
&OBST XB=18.20,18.30,9.60,10.10,3.10,3.20, SURF_ID='Wood Joists02'/ Rafter12
&OBST XB=18.20,18.30,10.00,10.50,3.00,3.10, SURF_ID='Wood Joists02'/ Rafter12
&OBST XB=18.20,18.30,10.400,10.90,2.90,3.00, SURF_ID='Wood Joists02'/ Rafter12
&OBST XB=18.20,18.30,10.80,11.30,2.80,2.90, SURF_ID='Wood Joists02'/ Rafter12
&OBST XB=18.20,18.30,11.20,12.80,2.70,2.80, SURF_ID='Wood Joists02'/ Rafter12
&OBST XB=18.20,18.30,0.50,7.60,2.50,2.70, SURF_ID='Wood Joists02'/ Rafter12
&OBST XB=17.60,17.70,0.50,7.60,2.50,2.70, SURF_ID='Wood Joists02'/ Rafter13
&OBST XB=17.60,17.70,0.80,1.30,2.70,2.80, SURF_ID='Wood Joists02'/ Rafter13
&OBST XB=17.60,17.70,1.20,1.70,2.80,2.90, SURF_ID='Wood Joists02'/ Rafter13
&OBST XB=17.60,17.70,1.60,2.10,2.90,3.00, SURF_ID='Wood Joists02'/ Rafter13
&OBST XB=17.60,17.70,2.00,2.50,3.00,3.10, SURF_ID='Wood Joists02'/ Rafter13
&OBST XB=17.60,17.70,2.40,2.90,3.10,3.20, SURF_ID='Wood Joists02'/ Rafter13
&OBST XB=17.60,17.70,2.80,3.30,3.20,3.30, SURF_ID='Wood Joists02'/ Rafter13
&OBST XB=17.60,17.70,3.20,3.70,3.30,3.40, SURF_ID='Wood Joists02'/ Rafter13
&OBST XB=17.60,17.70,3.60,4.10,3.40,3.50, SURF_ID='Wood Joists02'/ Rafter13
&OBST XB=17.60,17.70,4.00,4.50,3.50,3.60, SURF_ID='Wood Joists02'/ Rafter13
&OBST XB=17.60,17.70,4.40,4.90,3.60,3.70, SURF_ID='Wood Joists02'/ Rafter13
&OBST XB=17.60,17.70,4.80,5.30,3.70,3.80, SURF_ID='Wood Joists02'/ Rafter13
&OBST XB=17.60,17.70,5.20,5.70,3.80,3.90, SURF_ID='Wood Joists02'/ Rafter13
&OBST XB=17.60,17.70,5.60,6.20,3.90,4.00, SURF_ID='Wood Joists02'/ Rafter13
&OBST XB=17.60,17.70,6.30,6.90,3.90,4.00, SURF_ID='Wood Joists02'/ Rafter13
&OBST XB=17.60,17.70,6.80,7.30,3.80,3.90, SURF_ID='Wood Joists02'/ Rafter13
&OBST XB=17.60,17.70,7.20,7.70,3.70,3.80, SURF_ID='Wood Joists02'/ Rafter13
&OBST XB=17.60,17.70,7.60,8.10,3.60,3.70, SURF_ID='Wood Joists02'/ Rafter13
&OBST XB=17.60,17.70,8.00,8.50,3.50,3.60, SURF_ID='Wood Joists02'/ Rafter13
&OBST XB=17.60,17.70,8.40,8.90,3.40,3.50, SURF_ID='Wood Joists02'/ Rafter13
&OBST XB=17.60,17.70,8.80,9.30,3.30,3.40, SURF_ID='Wood Joists02'/ Rafter13
&OBST XB=17.60,17.70,9.20,9.70,3.20,3.30, SURF_ID='Wood Joists02'/ Rafter13
&OBST XB=17.60,17.70,9.60,10.10,3.10,3.20, SURF_ID='Wood Joists02'/ Rafter13
&OBST XB=17.60,17.70,10.00,10.50,3.00,3.10, SURF_ID='Wood Joists02'/ Rafter13
&OBST XB=17.60,17.70,10.400,10.90,2.90,3.00, SURF_ID='Wood Joists02'/ Rafter13
&OBST XB=17.60,17.70,10.80,11.30,2.80,2.90, SURF_ID='Wood Joists02'/ Rafter13
&OBST XB=17.60,17.70,11.20,12.80,2.70,2.80, SURF_ID='Wood Joists02'/ Rafter13
&OBST XB=17.00,17.10,0.50,7.60,2.50,2.70, SURF_ID='Wood Joists02'/ Rafter14
&OBST XB=17.00,17.10,0.80,1.30,2.70,2.80, SURF_ID='Wood Joists02'/ Rafter14
&OBST XB=17.00,17.10,1.20,1.70,2.80,2.90, SURF_ID='Wood Joists02'/ Rafter14
&OBST XB=17.00,17.10,1.60,2.10,2.90,3.00, SURF_ID='Wood Joists02'/ Rafter14
&OBST XB=17.00,17.10,2.00,2.50,3.00,3.10, SURF_ID='Wood Joists02'/ Rafter14
&OBST XB=17.00,17.10,2.40,2.90,3.10,3.20, SURF_ID='Wood Joists02'/ Rafter14
&OBST XB=17.00,17.10,2.80,3.30,3.20,3.30, SURF_ID='Wood Joists02'/ Rafter14
&OBST XB=17.00,17.10,3.20,3.70,3.30,3.40, SURF_ID='Wood Joists02'/ Rafter14
&OBST XB=17.00,17.10,3.60,4.10,3.40,3.50, SURF_ID='Wood Joists02'/ Rafter14
&OBST XB=17.00,17.10,4.00,4.50,3.50,3.60, SURF_ID='Wood Joists02'/ Rafter14
&OBST XB=17.00,17.10,4.40,4.90,3.60,3.70, SURF_ID='Wood Joists02'/ Rafter14
&OBST XB=17.00,17.10,4.80,5.30,3.70,3.80, SURF_ID='Wood Joists02'/ Rafter14
&OBST XB=17.00,17.10,5.20,5.70,3.80,3.90, SURF_ID='Wood Joists02'/ Rafter14
&OBST XB=17.00,17.10,5.60,6.20,3.90,4.00, SURF_ID='Wood Joists02'/ Rafter14
&OBST XB=17.00,17.10,6.30,6.90,3.90,4.00, SURF_ID='Wood Joists02'/ Rafter14
&OBST XB=17.00,17.10,6.80,7.30,3.80,3.90, SURF_ID='Wood Joists02'/ Rafter14
&OBST XB=17.00,17.10,7.20,7.70,3.70,3.80, SURF_ID='Wood Joists02'/ Rafter14
&OBST XB=17.00,17.10,7.60,8.10,3.60,3.70, SURF_ID='Wood Joists02'/ Rafter14
&OBST XB=17.00,17.10,8.00,8.50,3.50,3.60, SURF_ID='Wood Joists02'/ Rafter14
&OBST XB=17.00,17.10,8.40,8.90,3.40,3.50, SURF_ID='Wood Joists02'/ Rafter14
&OBST XB=17.00,17.10,8.80,9.30,3.30,3.40, SURF_ID='Wood Joists02'/ Rafter14
&OBST XB=17.00,17.10,9.20,9.70,3.20,3.30, SURF_ID='Wood Joists02'/ Rafter14
&OBST XB=17.00,17.10,9.60,10.10,3.10,3.20, SURF_ID='Wood Joists02'/ Rafter14
&OBST XB=17.00,17.10,10.00,10.50,3.00,3.10, SURF_ID='Wood Joists02'/ Rafter14
&OBST XB=17.00,17.10,10.400,10.90,2.90,3.00, SURF_ID='Wood Joists02'/ Rafter14
&OBST XB=17.00,17.10,10.80,11.30,2.80,2.90, SURF_ID='Wood Joists02'/ Rafter14
&OBST XB=17.00,17.10,11.20,12.80,2.70,2.80, SURF_ID='Wood Joists02'/ Rafter14
&OBST XB=16.40,16.50,0.50,7.60,2.50,2.70, SURF_ID='Wood Joists02'/ Rafter15
&OBST XB=16.40,16.50,0.80,1.30,2.70,2.80, SURF_ID='Wood Joists02'/ Rafter15
&OBST XB=16.40,16.50,1.20,1.70,2.80,2.90, SURF_ID='Wood Joists02'/ Rafter15
&OBST XB=16.40,16.50,1.60,2.10,2.90,3.00, SURF_ID='Wood Joists02'/ Rafter15
&OBST XB=16.40,16.50,2.00,2.50,3.00,3.10, SURF_ID='Wood Joists02'/ Rafter15
&OBST XB=16.40,16.50,2.40,2.90,3.10,3.20, SURF_ID='Wood Joists02'/ Rafter15
&OBST XB=16.40,16.50,2.80,3.30,3.20,3.30, SURF_ID='Wood Joists02'/ Rafter15
&OBST XB=16.40,16.50,3.20,3.70,3.30,3.40, SURF_ID='Wood Joists02'/ Rafter15
&OBST XB=16.40,16.50,3.60,4.10,3.40,3.50, SURF_ID='Wood Joists02'/ Rafter15
&OBST XB=16.40,16.50,4.00,4.50,3.50,3.60, SURF_ID='Wood Joists02'/ Rafter15
&OBST XB=16.40,16.50,4.40,4.90,3.60,3.70, SURF_ID='Wood Joists02'/ Rafter15
&OBST XB=16.40,16.50,4.80,5.30,3.70,3.80, SURF_ID='Wood Joists02'/ Rafter15
&OBST XB=16.40,16.50,5.20,5.70,3.80,3.90, SURF_ID='Wood Joists02'/ Rafter15
&OBST XB=16.40,16.50,5.60,6.20,3.90,4.00, SURF_ID='Wood Joists02'/ Rafter15
&OBST XB=16.40,16.50,6.30,6.90,3.90,4.00, SURF_ID='Wood Joists02'/ Rafter15
&OBST XB=16.40,16.50,6.80,7.30,3.80,3.90, SURF_ID='Wood Joists02'/ Rafter15
&OBST XB=16.40,16.50,7.20,7.70,3.70,3.80, SURF_ID='Wood Joists02'/ Rafter15
&OBST XB=16.40,16.50,7.60,8.10,3.60,3.70, SURF_ID='Wood Joists02'/ Rafter15
&OBST XB=16.40,16.50,8.00,8.50,3.50,3.60, SURF_ID='Wood Joists02'/ Rafter15
&OBST XB=16.40,16.50,8.40,8.90,3.40,3.50, SURF_ID='Wood Joists02'/ Rafter15
&OBST XB=16.40,16.50,8.80,9.30,3.30,3.40, SURF_ID='Wood Joists02'/ Rafter15
&OBST XB=16.40,16.50,9.20,9.70,3.20,3.30, SURF_ID='Wood Joists02'/ Rafter15
&OBST XB=16.40,16.50,9.60,10.10,3.10,3.20, SURF_ID='Wood Joists02'/ Rafter15
&OBST XB=16.40,16.50,10.00,10.50,3.00,3.10, SURF_ID='Wood Joists02'/ Rafter15
&OBST XB=16.40,16.50,10.400,10.90,2.90,3.00, SURF_ID='Wood Joists02'/ Rafter15
&OBST XB=16.40,16.50,10.80,11.30,2.80,2.90, SURF_ID='Wood Joists02'/ Rafter15
&OBST XB=16.40,16.50,11.20,12.80,2.70,2.80, SURF_ID='Wood Joists02'/ Rafter15
&OBST XB=15.80,15.90,0.50,7.60,2.50,2.70, SURF_ID='Wood Joists02'/ Rafter16
```

```
&OBST XB=15.80,15.90,0.80,1.30,2.70,2.80, SURF_ID='Wood Joists02'/ Rafter16
&OBST XB=15.80,15.90,1.20,1.70,2.80,2.90, SURF_ID='Wood Joists02'/ Rafter16
&OBST XB=15.80,15.90,1.60,2.10,2.90,3.00, SURF_ID='Wood Joists02'/ Rafter16
&OBST XB=15.80,15.90,2.00,2.50,3.00,3.10, SURF_ID='Wood Joists02'/ Rafter16
&OBST XB=15.80,15.90,2.40,2.90,3.10,3.20, SURF_ID='Wood Joists02'/ Rafter16
&OBST XB=15.80,15.90,2.80,3.30,3.20,3.30, SURF_ID='Wood Joists02'/ Rafter16
&OBST XB=15.80,15.90,3.20,3.70,3.30,3.40, SURF_ID='Wood Joists02'/ Rafter16
&OBST XB=15.80,15.90,3.60,4.10,3.40,3.50, SURF_ID='Wood Joists02'/ Rafter16
&OBST XB=15.80,15.90,4.00,4.50,3.50,3.60, SURF_ID='Wood Joists02'/ Rafter16
&OBST XB=15.80,15.90,4.40,4.90,3.60,3.70, SURF_ID='Wood Joists02'/ Rafter16
&OBST XB=15.80,15.90,4.80,5.30,3.70,3.80, SURF_ID='Wood Joists02'/ Rafter16
&OBST XB=15.80,15.90,5.20,5.70,3.80,3.90, SURF_ID='Wood Joists02'/ Rafter16
&OBST XB=15.80,15.90,5.60,6.20,3.90,4.00, SURF_ID='Wood Joists02'/ Rafter16
&OBST XB=15.80,15.90,6.30,6.90,3.90,4.00, SURF_ID='Wood Joists02'/ Rafter16
&OBST XB=15.80,15.90,6.80,7.30,3.80,3.90, SURF_ID='Wood Joists02'/ Rafter16
&OBST XB=15.80,15.90,7.20,7.70,3.70,3.80, SURF_ID='Wood Joists02'/ Rafter16
&OBST XB=15.80,15.90,7.60,8.10,3.60,3.70, SURF_ID='Wood Joists02'/ Rafter16
&OBST XB=15.80,15.90,8.00,8.50,3.50,3.60, SURF_ID='Wood Joists02'/ Rafter16
&OBST XB=15.80,15.90,8.40,8.90,3.40,3.50, SURF_ID='Wood Joists02'/ Rafter16
&OBST XB=15.80,15.90,8.80,9.30,3.30,3.40, SURF_ID='Wood Joists02'/ Rafter16
&OBST XB=15.80,15.90,9.20,9.70,3.20,3.30, SURF_ID='Wood Joists02'/ Rafter16
&OBST XB=15.80,15.90,9.60,10.10,3.10,3.20, SURF_ID='Wood Joists02'/ Rafter16
&OBST XB=15.80,15.90,10.00,10.50,3.00,3.10, SURF_ID='Wood Joists02'/ Rafter16
&OBST XB=15.80,15.90,10.400,10.90,2.90,3.00, SURF_ID='Wood Joists02'/ Rafter16
&OBST XB=15.80,15.90,10.80,11.30,2.80,2.90, SURF_ID='Wood Joists02'/ Rafter16
&OBST XB=15.80,15.90,11.20,12.80,2.70,2.80, SURF_ID='Wood Joists02'/ Rafter16
&OBST XB=15.20,15.30,0.50,12.10,2.50,2.70, SURF_ID='Wood Joists02'/ Rafter17
&OBST XB=15.20,15.30,0.80,1.30,2.70,2.80, SURF_ID='Wood Joists02'/ Rafter17
&OBST XB=15.20,15.30,1.20,1.70,2.80,2.90, SURF_ID='Wood Joists02'/ Rafter17
&OBST XB=15.20,15.30,1.60,2.10,2.90,3.00, SURF_ID='Wood Joists02'/ Rafter17
&OBST XB=15.20,15.30,2.00,2.50,3.00,3.10, SURF_ID='Wood Joists02'/ Rafter17
&OBST XB=15.20,15.30,2.40,2.90,3.10,3.20, SURF_ID='Wood Joists02'/ Rafter17
&OBST XB=15.20,15.30,2.80,3.30,3.20,3.30, SURF_ID='Wood Joists02'/ Rafter17
&OBST XB=15.20,15.30,3.20,3.70,3.30,3.40, SURF_ID='Wood Joists02'/ Rafter17
&OBST XB=15.20,15.30,3.60,4.10,3.40,3.50, SURF_ID='Wood Joists02'/ Rafter17
&OBST XB=15.20,15.30,4.00,4.50,3.50,3.60, SURF_ID='Wood Joists02'/ Rafter17
&OBST XB=15.20,15.30,4.40,4.90,3.60,3.70, SURF_ID='Wood Joists02'/ Rafter17
&OBST XB=15.20,15.30,4.80,5.30,3.70,3.80, SURF_ID='Wood Joists02'/ Rafter17
&OBST XB=15.20,15.30,5.20,5.70,3.80,3.90, SURF_ID='Wood Joists02'/ Rafter17
&OBST XB=15.20,15.30,5.60,6.20,3.90,4.00, SURF_ID='Wood Joists02'/ Rafter17
&OBST XB=15.20,15.30,6.30,6.90,3.90,4.00, SURF_ID='Wood Joists02'/ Rafter17
&OBST XB=15.20,15.30,6.80,7.30,3.80,3.90, SURF_ID='Wood Joists02'/ Rafter17
&OBST XB=15.20,15.30,7.20,7.70,3.70,3.80, SURF_ID='Wood Joists02'/ Rafter17
&OBST XB=15.20,15.30,7.60,8.10,3.60,3.70, SURF_ID='Wood Joists02'/ Rafter17
&OBST XB=15.20,15.30,8.00,8.50,3.50,3.60, SURF_ID='Wood Joists02'/ Rafter17
&OBST XB=15.20,15.30,8.40,8.90,3.40,3.50, SURF_ID='Wood Joists02'/ Rafter17
&OBST XB=15.20,15.30,8.80,9.30,3.30,3.40, SURF_ID='Wood Joists02'/ Rafter17
&OBST XB=15.20,15.30,9.20,9.70,3.20,3.30, SURF_ID='Wood Joists02'/ Rafter17
&OBST XB=15.20,15.30,9.60,10.10,3.10,3.20, SURF_ID='Wood Joists02'/ Rafter17
&OBST XB=15.20,15.30,10.00,10.50,3.00,3.10, SURF_ID='Wood Joists02'/ Rafter17
&OBST XB=15.20,15.30,10.400,10.90,2.90,3.00, SURF_ID='Wood Joists02'/ Rafter17
&OBST XB=15.20,15.30,10.80,11.30,2.80,2.90, SURF_ID='Wood Joists02'/ Rafter17
&OBST XB=15.20,15.30,11.20,11.70,2.70,2.80, SURF_ID='Wood Joists02'/ Rafter17
&OBST XB=14.60,14.70,0.50,14.90,2.50,2.70, SURF_ID='Wood Joists02'/ Rafter18
&OBST XB=14.60,14.70,0.80,1.30,2.70,2.80, SURF_ID='Wood Joists02'/ Rafter18
&OBST XB=14.60,14.70,1.20,1.70,2.80,2.90, SURF_ID='Wood Joists02'/ Rafter18
&OBST XB=14.60,14.70,1.60,2.10,2.90,3.00, SURF_ID='Wood Joists02'/ Rafter18
&OBST XB=14.60,14.70,2.00,2.50,3.00,3.10, SURF_ID='Wood Joists02'/ Rafter18
&OBST XB=14.60,14.70,2.40,2.90,3.10,3.20, SURF_ID='Wood Joists02'/ Rafter18
&OBST XB=14.60,14.70,2.80,3.30,3.20,3.30, SURF_ID='Wood Joists02'/ Rafter18
&OBST XB=14.60,14.70,3.20,3.70,3.30,3.40, SURF_ID='Wood Joists02'/ Rafter18
&OBST XB=14.60,14.70,3.60,4.10,3.40,3.50, SURF_ID='Wood Joists02'/ Rafter18
&OBST XB=14.60,14.70,4.00,4.50,3.50,3.60, SURF_ID='Wood Joists02'/ Rafter18
&OBST XB=14.60,14.70,4.40,4.90,3.60,3.70, SURF_ID='Wood Joists02'/ Rafter18
&OBST XB=14.60,14.70,4.80,5.30,3.70,3.80, SURF_ID='Wood Joists02'/ Rafter18
&OBST XB=14.60,14.70,5.20,5.70,3.80,3.90, SURF_ID='Wood Joists02'/ Rafter18
&OBST XB=14.60,14.70,5.60,6.20,3.90,4.00, SURF_ID='Wood Joists02'/ Rafter18
&OBST XB=14.60,14.70,6.30,6.90,3.90,4.00, SURF_ID='Wood Joists02'/ Rafter18
&OBST XB=14.60,14.70,6.80,7.30,3.80,3.90, SURF_ID='Wood Joists02'/ Rafter18
&OBST XB=14.60,14.70,7.20,7.70,3.70,3.80, SURF_ID='Wood Joists02'/ Rafter18
&OBST XB=14.60,14.70,7.60,8.10,3.60,3.70, SURF_ID='Wood Joists02'/ Rafter18
&OBST XB=14.60,14.70,8.00,8.50,3.50,3.60, SURF_ID='Wood Joists02'/ Rafter18
&OBST XB=14.60,14.70,8.40,8.90,3.40,3.50, SURF_ID='Wood Joists02'/ Rafter18
&OBST XB=14.60,14.70,8.80,9.30,3.30,3.40, SURF_ID='Wood Joists02'/ Rafter18
&OBST XB=14.60,14.70,9.20,9.70,3.20,3.30, SURF_ID='Wood Joists02'/ Rafter18
&OBST XB=14.60,14.70,9.60,10.10,3.10,3.20, SURF_ID='Wood Joists02'/ Rafter18
&OBST XB=14.60,14.70,10.00,10.50,3.00,3.10, SURF_ID='Wood Joists02'/ Rafter18
&OBST XB=14.60,14.70,10.400,10.90,2.90,3.00, SURF_ID='Wood Joists02'/ Rafter18
&OBST XB=14.60,14.70,10.80,11.30,2.80,2.90, SURF_ID='Wood Joists02'/ Rafter18
&OBST XB=14.60,14.70,11.20,11.70,2.70,2.80, SURF_ID='Wood Joists02'/ Rafter18
&OBST XB=14.00,14.10,0.50,14.90,2.50,2.70, SURF_ID='Wood Joists02'/ Rafter19
&OBST XB=14.00,14.10,0.80,1.30,2.70,2.80, SURF_ID='Wood Joists02'/ Rafter19
&OBST XB=14.00,14.10,1.20,1.70,2.80,2.90, SURF_ID='Wood Joists02'/ Rafter19
&OBST XB=14.00,14.10,1.60,2.10,2.90,3.00, SURF_ID='Wood Joists02'/ Rafter19
&OBST XB=14.00,14.10,2.00,2.50,3.00,3.10, SURF_ID='Wood Joists02'/ Rafter19
&OBST XB=14.00,14.10,2.40,2.90,3.10,3.20, SURF_ID='Wood Joists02'/ Rafter19
&OBST XB=14.00,14.10,2.80,3.30,3.20,3.30, SURF_ID='Wood Joists02'/ Rafter19
&OBST XB=14.00,14.10,3.20,3.70,3.30,3.40, SURF_ID='Wood Joists02'/ Rafter19
&OBST XB=14.00,14.10,3.60,4.10,3.40,3.50, SURF_ID='Wood Joists02'/ Rafter19
&OBST XB=14.00,14.10,4.00,4.50,3.50,3.60, SURF_ID='Wood Joists02'/ Rafter19
&OBST XB=14.00,14.10,4.40,4.90,3.60,3.70, SURF_ID='Wood Joists02'/ Rafter19
&OBST XB=14.00,14.10,4.80,5.30,3.70,3.80, SURF_ID='Wood Joists02'/ Rafter19
&OBST XB=14.00,14.10,5.20,5.70,3.80,3.90, SURF_ID='Wood Joists02'/ Rafter19
&OBST XB=14.00,14.10,5.60,6.20,3.90,4.00, SURF_ID='Wood Joists02'/ Rafter19
&OBST XB=14.00,14.10,6.30,6.90,3.90,4.00, SURF_ID='Wood Joists02'/ Rafter19
```

```
&OBST XB=14.00,14.10,6.80,7.30,3.80,3.90, SURF_ID='Wood Joists02'/ Rafter19
&OBST XB=14.00,14.10,7.20,7.70,3.70,3.80, SURF_ID='Wood Joists02'/ Rafter19
&OBST XB=14.00,14.10,7.60,8.10,3.60,3.70, SURF_ID='Wood Joists02'/ Rafter19
&OBST XB=14.00,14.10,8.00,8.50,3.50,3.60, SURF_ID='Wood Joists02'/ Rafter19
&OBST XB=14.00,14.10,8.40,8.90,3.40,3.50, SURF_ID='Wood Joists02'/ Rafter19
&OBST XB=14.00,14.10,8.80,9.30,3.30,3.40, SURF_ID='Wood Joists02'/ Rafter19
&OBST XB=14.00,14.10,9.20,9.70,3.20,3.30, SURF_ID='Wood Joists02'/ Rafter19
&OBST XB=14.00,14.10,9.60,10.10,3.10,3.20, SURF_ID='Wood Joists02'/ Rafter19
&OBST XB=14.00,14.10,10.00,10.50,3.00,3.10, SURF_ID='Wood Joists02'/ Rafter19
&OBST XB=14.00,14.10,10.400,10.90,2.90,3.00, SURF_ID='Wood Joists02'/ Rafter19
&OBST XB=14.00,14.10,10.80,11.30,2.80,2.90, SURF_ID='Wood Joists02'/ Rafter19
&OBST XB=14.00,14.10,11.20,11.70,2.70,2.80, SURF_ID='Wood Joists02'/ Rafter19
&OBST XB=13.40,13.50,0.50,14.90,2.50,2.70, SURF_ID='Wood Joists02'/ Rafter20
&OBST XB=13.40,13.50,0.80,1.30,2.70,2.80, SURF_ID='Wood Joists02'/ Rafter20
&OBST XB=13.40,13.50,1.20,1.70,2.80,2.90, SURF_ID='Wood Joists02'/ Rafter20
&OBST XB=13.40,13.50,1.60,2.10,2.90,3.00, SURF_ID='Wood Joists02'/ Rafter20
&OBST XB=13.40,13.50,2.00,2.50,3.00,3.10, SURF_ID='Wood Joists02'/ Rafter20
&OBST XB=13.40,13.50,2.40,2.90,3.10,3.20, SURF_ID='Wood Joists02'/ Rafter20
&OBST XB=13.40,13.50,2.80,3.30,3.20,3.30, SURF_ID='Wood Joists02'/ Rafter20
&OBST XB=13.40,13.50,3.20,3.70,3.30,3.40, SURF_ID='Wood Joists02'/ Rafter20
&OBST XB=13.40,13.50,3.60,4.10,3.40,3.50, SURF_ID='Wood Joists02'/ Rafter20
&OBST XB=13.40,13.50,4.00,4.50,3.50,3.60, SURF_ID='Wood Joists02'/ Rafter20
&OBST XB=13.40,13.50,4.40,4.90,3.60,3.70, SURF_ID='Wood Joists02'/ Rafter20
&OBST XB=13.40,13.50,4.80,5.30,3.70,3.80, SURF_ID='Wood Joists02'/ Rafter20
&OBST XB=13.40,13.50,5.20,5.70,3.80,3.90, SURF_ID='Wood Joists02'/ Rafter20
&OBST XB=13.40,13.50,5.60,6.20,3.90,4.00, SURF_ID='Wood Joists02'/ Rafter20
&OBST XB=13.40,13.50,6.30,6.90,3.90,4.00, SURF_ID='Wood Joists02'/ Rafter20
&OBST XB=13.40,13.50,6.80,7.30,3.80,3.90, SURF_ID='Wood Joists02'/ Rafter20
&OBST XB=13.40,13.50,7.20,7.70,3.70,3.80, SURF_ID='Wood Joists02'/ Rafter20
&OBST XB=13.40,13.50,7.60,8.10,3.60,3.70, SURF_ID='Wood Joists02'/ Rafter20
&OBST XB=13.40,13.50,8.00,8.50,3.50,3.60, SURF_ID='Wood Joists02'/ Rafter20
&OBST XB=13.40,13.50,8.40,8.90,3.40,3.50, SURF_ID='Wood Joists02'/ Rafter20
&OBST XB=13.40,13.50,8.80,9.30,3.30,3.40, SURF_ID='Wood Joists02'/ Rafter20
&OBST XB=13.40,13.50,9.20,9.70,3.20,3.30, SURF_ID='Wood Joists02'/ Rafter20
&OBST XB=13.40,13.50,9.60,10.10,3.10,3.20, SURF_ID='Wood Joists02'/ Rafter20
&OBST XB=13.40,13.50,10.00,10.50,3.00,3.10, SURF_ID='Wood Joists02'/ Rafter20
&OBST XB=13.40,13.50,10.400,10.90,2.90,3.00, SURF_ID='Wood Joists02'/ Rafter20
&OBST XB=13.40,13.50,10.80,11.30,2.80,2.90, SURF_ID='Wood Joists02'/ Rafter20
&OBST XB=13.40,13.50,11.20,11.70,2.70,2.80, SURF_ID='Wood Joists02'/ Rafter20
&OBST XB=12.80,12.90,0.50,14.90,2.50,2.70, SURF_ID='Wood Joists02'/ Rafter21
&OBST XB=12.80,12.90,0.80,1.30,2.70,2.80, SURF_ID='Wood Joists02'/ Rafter21
&OBST XB=12.80,12.90,1.20,1.70,2.80,2.90, SURF_ID='Wood Joists02'/ Rafter21
&OBST XB=12.80,12.90,1.60,2.10,2.90,3.00, SURF_ID='Wood Joists02'/ Rafter21
&OBST XB=12.80,12.90,2.00,2.50,3.00,3.10, SURF_ID='Wood Joists02'/ Rafter21
&OBST XB=12.80,12.90,2.40,2.90,3.10,3.20, SURF_ID='Wood Joists02'/ Rafter21
&OBST XB=12.80,12.90,2.80,3.30,3.20,3.30, SURF_ID='Wood Joists02'/ Rafter21
&OBST XB=12.80,12.90,3.20,3.70,3.30,3.40, SURF_ID='Wood Joists02'/ Rafter21
&OBST XB=12.80,12.90,3.60,4.10,3.40,3.50, SURF_ID='Wood Joists02'/ Rafter21
&OBST XB=12.80,12.90,4.00,4.50,3.50,3.60, SURF_ID='Wood Joists02'/ Rafter21
&OBST XB=12.80,12.90,4.40,4.90,3.60,3.70, SURF_ID='Wood Joists02'/ Rafter21
&OBST XB=12.80,12.90,4.80,5.30,3.70,3.80, SURF_ID='Wood Joists02'/ Rafter21
&OBST XB=12.80,12.90,5.20,5.70,3.80,3.90, SURF_ID='Wood Joists02'/ Rafter21
&OBST XB=12.80,12.90,5.60,6.20,3.90,4.00, SURF_ID='Wood Joists02'/ Rafter21
&OBST XB=12.80,12.90,6.30,6.90,3.90,4.00, SURF_ID='Wood Joists02'/ Rafter21
&OBST XB=12.80,12.90,6.80,7.30,3.80,3.90, SURF_ID='Wood Joists02'/ Rafter21
&OBST XB=12.80,12.90,7.20,7.70,3.70,3.80, SURF_ID='Wood Joists02'/ Rafter21
&OBST XB=12.80,12.90,7.60,8.10,3.60,3.70, SURF_ID='Wood Joists02'/ Rafter21
&OBST XB=12.80,12.90,8.00,8.50,3.50,3.60, SURF_ID='Wood Joists02'/ Rafter21
&OBST XB=12.80,12.90,8.40,8.90,3.40,3.50, SURF_ID='Wood Joists02'/ Rafter21
&OBST XB=12.80,12.90,8.80,9.30,3.30,3.40, SURF_ID='Wood Joists02'/ Rafter21
&OBST XB=12.80,12.90,9.20,9.70,3.20,3.30, SURF_ID='Wood Joists02'/ Rafter21
&OBST XB=12.80,12.90,9.60,10.10,3.10,3.20, SURF_ID='Wood Joists02'/ Rafter21
&OBST XB=12.80,12.90,10.00,10.50,3.00,3.10, SURF_ID='Wood Joists02'/ Rafter21
&OBST XB=12.80,12.90,10.400,10.90,2.90,3.00, SURF_ID='Wood Joists02'/ Rafter21
&OBST XB=12.80,12.90,10.80,11.30,2.80,2.90, SURF_ID='Wood Joists02'/ Rafter21
&OBST XB=12.80,12.90,11.20,11.70,2.70,2.80, SURF_ID='Wood Joists02'/ Rafter21
&OBST XB=12.20,12.30,0.50,14.90,2.50,2.70, SURF_ID='Wood Joists02'/ Rafter22
&OBST XB=12.20,12.30,0.80,1.30,2.70,2.80, SURF_ID='Wood Joists02'/ Rafter22
&OBST XB=12.20,12.30,1.20,1.70,2.80,2.90, SURF_ID='Wood Joists02'/ Rafter22
&OBST XB=12.20,12.30,1.60,2.10,2.90,3.00, SURF_ID='Wood Joists02'/ Rafter22
&OBST XB=12.20,12.30,2.00,2.50,3.00,3.10, SURF_ID='Wood Joists02'/ Rafter22
&OBST XB=12.20,12.30,2.40,2.90,3.10,3.20, SURF_ID='Wood Joists02'/ Rafter22
&OBST XB=12.20,12.30,2.80,3.30,3.20,3.30, SURF_ID='Wood Joists02'/ Rafter22
&OBST XB=12.20,12.30,3.20,3.70,3.30,3.40, SURF_ID='Wood Joists02'/ Rafter22
&OBST XB=12.20,12.30,3.60,4.10,3.40,3.50, SURF_ID='Wood Joists02'/ Rafter22
&OBST XB=12.20,12.30,4.00,4.50,3.50,3.60, SURF_ID='Wood Joists02'/ Rafter22
&OBST XB=12.20,12.30,4.40,4.90,3.60,3.70, SURF_ID='Wood Joists02'/ Rafter22
&OBST XB=12.20,12.30,4.80,5.30,3.70,3.80, SURF_ID='Wood Joists02'/ Rafter22
&OBST XB=12.20,12.30,5.20,5.70,3.80,3.90, SURF_ID='Wood Joists02'/ Rafter22
&OBST XB=12.20,12.30,5.60,6.20,3.90,4.00, SURF_ID='Wood Joists02'/ Rafter22
&OBST XB=12.20,12.30,6.30,6.90,3.90,4.00, SURF_ID='Wood Joists02'/ Rafter22
&OBST XB=12.20,12.30,6.80,7.30,3.80,3.90, SURF_ID='Wood Joists02'/ Rafter22
&OBST XB=12.20,12.30,7.20,7.70,3.70,3.80, SURF_ID='Wood Joists02'/ Rafter22
&OBST XB=12.20,12.30,7.60,8.10,3.60,3.70, SURF_ID='Wood Joists02'/ Rafter22
&OBST XB=12.20,12.30,8.00,8.50,3.50,3.60, SURF_ID='Wood Joists02'/ Rafter22
&OBST XB=12.20,12.30,8.40,8.90,3.40,3.50, SURF_ID='Wood Joists02'/ Rafter22
&OBST XB=12.20,12.30,8.80,9.30,3.30,3.40, SURF_ID='Wood Joists02'/ Rafter22
&OBST XB=12.20,12.30,9.20,9.70,3.20,3.30, SURF_ID='Wood Joists02'/ Rafter22
&OBST XB=12.20,12.30,9.60,10.10,3.10,3.20, SURF_ID='Wood Joists02'/ Rafter22
&OBST XB=12.20,12.30,10.00,10.50,3.00,3.10, SURF_ID='Wood Joists02'/ Rafter22
&OBST XB=12.20,12.30,10.400,10.90,2.90,3.00, SURF_ID='Wood Joists02'/ Rafter22
&OBST XB=12.20,12.30,10.80,11.30,2.80,2.90, SURF_ID='Wood Joists02'/ Rafter22
&OBST XB=12.20,12.30,11.20,11.70,2.70,2.80, SURF_ID='Wood Joists02'/ Rafter22
&OBST XB=11.60,11.70,0.50,14.90,2.50,2.70, SURF_ID='Wood Joists02'/ Rafter23
&OBST XB=11.60,11.70,0.80,1.30,2.70,2.80, SURF_ID='Wood Joists02'/ Rafter23
```

```
&OBST XB=11.60,11.70,1.20,1.70,2.80,2.90, SURF_ID='Wood Joists02'/ Rafter23
&OBST XB=11.60,11.70,1.60,2.10,2.90,3.00, SURF_ID='Wood Joists02'/ Rafter23
&OBST XB=11.60,11.70,2.00,2.50,3.00,3.10, SURF_ID='Wood Joists02'/ Rafter23
&OBST XB=11.60,11.70,2.40,2.90,3.10,3.20, SURF_ID='Wood Joists02'/ Rafter23
&OBST XB=11.60,11.70,2.80,3.30,3.20,3.30, SURF_ID='Wood Joists02'/ Rafter23
&OBST XB=11.60,11.70,3.20,3.70,3.30,3.40, SURF_ID='Wood Joists02'/ Rafter23
&OBST XB=11.60,11.70,3.60,4.10,3.40,3.50, SURF_ID='Wood Joists02'/ Rafter23
&OBST XB=11.60,11.70,4.00,4.50,3.50,3.60, SURF_ID='Wood Joists02'/ Rafter23
&OBST XB=11.60,11.70,4.40,4.90,3.60,3.70, SURF_ID='Wood Joists02'/ Rafter23
&OBST XB=11.60,11.70,4.80,5.30,3.70,3.80, SURF_ID='Wood Joists02'/ Rafter23
&OBST XB=11.60,11.70,5.20,5.70,3.80,3.90, SURF_ID='Wood Joists02'/ Rafter23
&OBST XB=11.60,11.70,5.60,6.20,3.90,4.00, SURF_ID='Wood Joists02'/ Rafter23
&OBST XB=11.60,11.70,6.30,6.90,3.90,4.00, SURF_ID='Wood Joists02'/ Rafter23
&OBST XB=11.60,11.70,6.80,7.30,3.80,3.90, SURF_ID='Wood Joists02'/ Rafter23
&OBST XB=11.60,11.70,7.20,7.70,3.70,3.80, SURF_ID='Wood Joists02'/ Rafter23
&OBST XB=11.60,11.70,7.60,8.10,3.60,3.70, SURF_ID='Wood Joists02'/ Rafter23
&OBST XB=11.60,11.70,8.00,8.50,3.50,3.60, SURF_ID='Wood Joists02'/ Rafter23
&OBST XB=11.60,11.70,8.40,8.90,3.40,3.50, SURF_ID='Wood Joists02'/ Rafter23
&OBST XB=11.60,11.70,8.80,9.30,3.30,3.40, SURF_ID='Wood Joists02'/ Rafter23
&OBST XB=11.60,11.70,9.20,9.70,3.20,3.30, SURF_ID='Wood Joists02'/ Rafter23
&OBST XB=11.60,11.70,9.60,10.10,3.10,3.20, SURF_ID='Wood Joists02'/ Rafter23
&OBST XB=11.60,11.70,10.00,10.50,3.00,3.10, SURF_ID='Wood Joists02'/ Rafter23
&OBST XB=11.60,11.70,10.400,10.90,2.90,3.00, SURF_ID='Wood Joists02'/ Rafter23
&OBST XB=11.60,11.70,10.80,11.30,2.80,2.90, SURF_ID='Wood Joists02'/ Rafter23
&OBST XB=11.60,11.70,11.20,11.70,2.70,2.80, SURF_ID='Wood Joists02'/ Rafter23
&OBST XB=11.00,11.10,0.50,14.90,2.50,2.70, SURF_ID='Wood Joists02'/ Rafter24
&OBST XB=11.00,11.10,0.80,1.30,2.70,2.80, SURF_ID='Wood Joists02'/ Rafter24
&OBST XB=11.00,11.10,1.20,1.70,2.80,2.90, SURF_ID='Wood Joists02'/ Rafter24
&OBST XB=11.00,11.10,1.60,2.10,2.90,3.00, SURF_ID='Wood Joists02'/ Rafter24
&OBST XB=11.00,11.10,2.00,2.50,3.00,3.10, SURF_ID='Wood Joists02'/ Rafter24
&OBST XB=11.00,11.10,2.40,2.90,3.10,3.20, SURF_ID='Wood Joists02'/ Rafter24
&OBST XB=11.00,11.10,2.80,3.30,3.20,3.30, SURF_ID='Wood Joists02'/ Rafter24
&OBST XB=11.00,11.10,3.20,3.70,3.30,3.40, SURF_ID='Wood Joists02'/ Rafter24
&OBST XB=11.00,11.10,3.60,4.10,3.40,3.50, SURF_ID='Wood Joists02'/ Rafter24
&OBST XB=11.00,11.10,4.00,4.50,3.50,3.60, SURF_ID='Wood Joists02'/ Rafter24
&OBST XB=11.00,11.10,4.40,4.90,3.60,3.70, SURF_ID='Wood Joists02'/ Rafter24
&OBST XB=11.00,11.10,4.80,5.30,3.70,3.80, SURF_ID='Wood Joists02'/ Rafter24
&OBST XB=11.00,11.10,5.20,5.70,3.80,3.90, SURF_ID='Wood Joists02'/ Rafter24
&OBST XB=11.00,11.10,5.60,6.20,3.90,4.00, SURF_ID='Wood Joists02'/ Rafter24
&OBST XB=11.00,11.10,6.30,6.90,3.90,4.00, SURF_ID='Wood Joists02'/ Rafter24
&OBST XB=11.00,11.10,6.80,7.30,3.80,3.90, SURF_ID='Wood Joists02'/ Rafter24
&OBST XB=11.00,11.10,7.20,7.70,3.70,3.80, SURF_ID='Wood Joists02'/ Rafter24
&OBST XB=11.00,11.10,7.60,8.10,3.60,3.70, SURF_ID='Wood Joists02'/ Rafter24
&OBST XB=11.00,11.10,8.00,8.50,3.50,3.60, SURF_ID='Wood Joists02'/ Rafter24
&OBST XB=11.00,11.10,8.40,8.90,3.40,3.50, SURF_ID='Wood Joists02'/ Rafter24
&OBST XB=11.00,11.10,8.80,9.30,3.30,3.40, SURF_ID='Wood Joists02'/ Rafter24
&OBST XB=11.00,11.10,9.20,9.70,3.20,3.30, SURF_ID='Wood Joists02'/ Rafter24
&OBST XB=11.00,11.10,9.60,10.10,3.10,3.20, SURF_ID='Wood Joists02'/ Rafter24
&OBST XB=11.00,11.10,10.00,10.50,3.00,3.10, SURF_ID='Wood Joists02'/ Rafter24
&OBST XB=11.00,11.10,10.400,10.90,2.90,3.00, SURF_ID='Wood Joists02'/ Rafter24
&OBST XB=11.00,11.10,10.80,11.30,2.80,2.90, SURF_ID='Wood Joists02'/ Rafter24
&OBST XB=11.00,11.10,11.20,11.70,2.70,2.80, SURF_ID='Wood Joists02'/ Rafter24
&OBST XB=10.400,10.50,0.50,14.90,2.50,2.70, SURF_ID='Wood Joists02'/ Rafter25
&OBST XB=10.400,10.50,0.80,1.30,2.70,2.80, SURF_ID='Wood Joists02'/ Rafter25
&OBST XB=10.400,10.50,1.20,1.70,2.80,2.90, SURF_ID='Wood Joists02'/ Rafter25
&OBST XB=10.400,10.50,1.60,2.10,2.90,3.00, SURF_ID='Wood Joists02'/ Rafter25
&OBST XB=10.400,10.50,2.00,2.50,3.00,3.10, SURF_ID='Wood Joists02'/ Rafter25
&OBST XB=10.400,10.50,2.40,2.90,3.10,3.20, SURF_ID='Wood Joists02'/ Rafter25
&OBST XB=10.400,10.50,2.80,3.30,3.20,3.30, SURF_ID='Wood Joists02'/ Rafter25
&OBST XB=10.400,10.50,3.20,3.70,3.30,3.40, SURF_ID='Wood Joists02'/ Rafter25
&OBST XB=10.400,10.50,3.60,4.10,3.40,3.50, SURF_ID='Wood Joists02'/ Rafter25
&OBST XB=10.400,10.50,4.00,4.50,3.50,3.60, SURF_ID='Wood Joists02'/ Rafter25
&OBST XB=10.400,10.50,4.40,4.90,3.60,3.70, SURF_ID='Wood Joists02'/ Rafter25
&OBST XB=10.400,10.50,4.80,5.30,3.70,3.80, SURF_ID='Wood Joists02'/ Rafter25
&OBST XB=10.400,10.50,5.20,5.70,3.80,3.90, SURF_ID='Wood Joists02'/ Rafter25
&OBST XB=10.400,10.50,5.60,6.20,3.90,4.00, SURF_ID='Wood Joists02'/ Rafter25
&OBST XB=10.400,10.50,6.30,6.90,3.90,4.00, SURF_ID='Wood Joists02'/ Rafter25
&OBST XB=10.400,10.50,6.80,7.30,3.80,3.90, SURF_ID='Wood Joists02'/ Rafter25
&OBST XB=10.400,10.50,7.20,7.70,3.70,3.80, SURF_ID='Wood Joists02'/ Rafter25
&OBST XB=10.400,10.50,7.60,8.10,3.60,3.70, SURF_ID='Wood Joists02'/ Rafter25
&OBST XB=10.400,10.50,8.00,8.50,3.50,3.60, SURF_ID='Wood Joists02'/ Rafter25
&OBST XB=10.400,10.50,8.40,8.90,3.40,3.50, SURF_ID='Wood Joists02'/ Rafter25
&OBST XB=10.400,10.50,8.80,9.30,3.30,3.40, SURF_ID='Wood Joists02'/ Rafter25
&OBST XB=10.400,10.50,9.20,9.70,3.20,3.30, SURF_ID='Wood Joists02'/ Rafter25
&OBST XB=10.400,10.50,9.60,10.10,3.10,3.20, SURF_ID='Wood Joists02'/ Rafter25
&OBST XB=10.400,10.50,10.00,10.50,3.00,3.10, SURF_ID='Wood Joists02'/ Rafter25
&OBST XB=10.400,10.50,10.400,10.90,2.90,3.00, SURF_ID='Wood Joists02'/ Rafter25
&OBST XB=10.400,10.50,10.80,11.30,2.80,2.90, SURF_ID='Wood Joists02'/ Rafter25
&OBST XB=10.400,10.50,11.20,11.70,2.70,2.80, SURF_ID='Wood Joists02'/ Rafter25
&OBST XB=9.80,9.90,0.50,14.90,2.50,2.70, SURF_ID='Wood Joists02'/ Rafter26
&OBST XB=9.80,9.90,0.80,1.30,2.70,2.80, SURF_ID='Wood Joists02'/ Rafter26
&OBST XB=9.80,9.90,1.20,1.70,2.80,2.90, SURF_ID='Wood Joists02'/ Rafter26
&OBST XB=9.80,9.90,1.60,2.10,2.90,3.00, SURF_ID='Wood Joists02'/ Rafter26
&OBST XB=9.80,9.90,2.00,2.50,3.00,3.10, SURF_ID='Wood Joists02'/ Rafter26
&OBST XB=9.80,9.90,2.40,2.90,3.10,3.20, SURF_ID='Wood Joists02'/ Rafter26
&OBST XB=9.80,9.90,2.80,3.30,3.20,3.30, SURF_ID='Wood Joists02'/ Rafter26
&OBST XB=9.80,9.90,3.20,3.70,3.30,3.40, SURF_ID='Wood Joists02'/ Rafter26
&OBST XB=9.80,9.90,3.60,4.10,3.40,3.50, SURF_ID='Wood Joists02'/ Rafter26
&OBST XB=9.80,9.90,4.00,4.50,3.50,3.60, SURF_ID='Wood Joists02'/ Rafter26
&OBST XB=9.80,9.90,4.40,4.90,3.60,3.70, SURF_ID='Wood Joists02'/ Rafter26
&OBST XB=9.80,9.90,4.80,5.30,3.70,3.80, SURF_ID='Wood Joists02'/ Rafter26
&OBST XB=9.80,9.90,5.20,5.70,3.80,3.90, SURF_ID='Wood Joists02'/ Rafter26
&OBST XB=9.80,9.90,5.60,6.20,3.90,4.00, SURF_ID='Wood Joists02'/ Rafter26
&OBST XB=9.80,9.90,6.30,6.90,3.90,4.00, SURF_ID='Wood Joists02'/ Rafter26
&OBST XB=9.80,9.90,6.80,7.30,3.80,3.90, SURF_ID='Wood Joists02'/ Rafter26
```

65

```
&OBST XB=9.80,9.90,7.20,7.70,3.70,3.80, SURF_ID='Wood Joists02'/ Rafter26
&OBST XB=9.80,9.90,7.60,8.10,3.60,3.70, SURF_ID='Wood Joists02'/ Rafter26
&OBST XB=9.80,9.90,8.00,8.50,3.50,3.60, SURF_ID='Wood Joists02'/ Rafter26
&OBST XB=9.80,9.90,8.40,8.90,3.40,3.50, SURF_ID='Wood Joists02'/ Rafter26
&OBST XB=9.80,9.90,8.80,9.30,3.30,3.40, SURF_ID='Wood Joists02'/ Rafter26
&OBST XB=9.80,9.90,9.20,9.70,3.20,3.30, SURF_ID='Wood Joists02'/ Rafter26
&OBST XB=9.80,9.90,9.60,10.10,3.10,3.20, SURF_ID='Wood Joists02'/ Rafter26
&OBST XB=9.80,9.90,10.00,10.50,3.00,3.10, SURF_ID='Wood Joists02'/ Rafter26
&OBST XB=9.80,9.90,10.400,10.90,2.90,3.00, SURF_ID='Wood Joists02'/ Rafter26
&OBST XB=9.80,9.90,10.80,11.30,2.80,2.90, SURF_ID='Wood Joists02'/ Rafter26
&OBST XB=9.80,9.90,11.20,11.70,2.70,2.80, SURF_ID='Wood Joists02'/ Rafter26
&OBST XB=9.20,9.30,0.50,14.90,2.50,2.70, SURF_ID='Wood Joists02'/ Rafter27
&OBST XB=9.20,9.30,0.80,1.30,2.70,2.80, SURF_ID='Wood Joists02'/ Rafter27
&OBST XB=9.20,9.30,1.20,1.70,2.80,2.90, SURF_ID='Wood Joists02'/ Rafter27
&OBST XB=9.20,9.30,1.60,2.10,2.90,3.00, SURF_ID='Wood Joists02'/ Rafter27
&OBST XB=9.20,9.30,2.00,2.50,3.00,3.10, SURF_ID='Wood Joists02'/ Rafter27
&OBST XB=9.20,9.30,2.40,2.90,3.10,3.20, SURF_ID='Wood Joists02'/ Rafter27
&OBST XB=9.20,9.30,2.80,3.30,3.20,3.30, SURF_ID='Wood Joists02'/ Rafter27
&OBST XB=9.20,9.30,3.20,3.70,3.30,3.40, SURF_ID='Wood Joists02'/ Rafter27
&OBST XB=9.20,9.30,3.60,4.10,3.40,3.50, SURF_ID='Wood Joists02'/ Rafter27
&OBST XB=9.20,9.30,4.00,4.50,3.50,3.60, SURF_ID='Wood Joists02'/ Rafter27
&OBST XB=9.20,9.30,4.40,4.90,3.60,3.70, SURF_ID='Wood Joists02'/ Rafter27
&OBST XB=9.20,9.30,4.80,5.30,3.70,3.80, SURF_ID='Wood Joists02'/ Rafter27
&OBST XB=9.20,9.30,5.20,5.70,3.80,3.90, SURF_ID='Wood Joists02'/ Rafter27
&OBST XB=9.20,9.30,5.60,6.20,3.90,4.00, SURF_ID='Wood Joists02'/ Rafter27
&OBST XB=9.20,9.30,6.30,6.90,3.90,4.00, SURF_ID='Wood Joists02'/ Rafter27
&OBST XB=9.20,9.30,6.80,7.30,3.80,3.90, SURF_ID='Wood Joists02'/ Rafter27
&OBST XB=9.20,9.30,7.20,7.70,3.70,3.80, SURF_ID='Wood Joists02'/ Rafter27
&OBST XB=9.20,9.30,7.60,8.10,3.60,3.70, SURF_ID='Wood Joists02'/ Rafter27
&OBST XB=9.20,9.30,8.00,8.50,3.50,3.60, SURF_ID='Wood Joists02'/ Rafter27
&OBST XB=9.20,9.30,8.40,8.90,3.40,3.50, SURF_ID='Wood Joists02'/ Rafter27
&OBST XB=9.20,9.30,8.80,9.30,3.30,3.40, SURF_ID='Wood Joists02'/ Rafter27
&OBST XB=9.20,9.30,9.20,9.70,3.20,3.30, SURF_ID='Wood Joists02'/ Rafter27
&OBST XB=9.20,9.30,9.60,10.10,3.10,3.20, SURF_ID='Wood Joists02'/ Rafter27
&OBST XB=9.20,9.30,10.00,10.50,3.00,3.10, SURF_ID='Wood Joists02'/ Rafter27
&OBST XB=9.20,9.30,10.400,10.90,2.90,3.00, SURF_ID='Wood Joists02'/ Rafter27
&OBST XB=9.20,9.30,10.80,11.30,2.80,2.90, SURF_ID='Wood Joists02'/ Rafter27
&OBST XB=9.20,9.30,11.20,11.70,2.70,2.80, SURF_ID='Wood Joists02'/ Rafter27
&OBST XB=8.60,8.70,0.50,14.90,2.50,2.70, SURF_ID='Wood Joists02'/ Rafter28
&OBST XB=8.60,8.70,0.80,1.30,2.70,2.80, SURF_ID='Wood Joists02'/ Rafter28
&OBST XB=8.60,8.70,1.20,1.70,2.80,2.90, SURF_ID='Wood Joists02'/ Rafter28
&OBST XB=8.60,8.70,1.60,2.10,2.90,3.00, SURF_ID='Wood Joists02'/ Rafter28
&OBST XB=8.60,8.70,2.00,2.50,3.00,3.10, SURF_ID='Wood Joists02'/ Rafter28
&OBST XB=8.60,8.70,2.40,2.90,3.10,3.20, SURF_ID='Wood Joists02'/ Rafter28
&OBST XB=8.60,8.70,2.80,3.30,3.20,3.30, SURF_ID='Wood Joists02'/ Rafter28
&OBST XB=8.60,8.70,3.20,3.70,3.30,3.40, SURF_ID='Wood Joists02'/ Rafter28
&OBST XB=8.60,8.70,3.60,4.10,3.40,3.50, SURF_ID='Wood Joists02'/ Rafter28
&OBST XB=8.60,8.70,4.00,4.50,3.50,3.60, SURF_ID='Wood Joists02'/ Rafter28
&OBST XB=8.60,8.70,4.40,4.90,3.60,3.70, SURF_ID='Wood Joists02'/ Rafter28
&OBST XB=8.60,8.70,4.80,5.30,3.70,3.80, SURF_ID='Wood Joists02'/ Rafter28
&OBST XB=8.60,8.70,5.20,5.70,3.80,3.90, SURF_ID='Wood Joists02'/ Rafter28
&OBST XB=8.60,8.70,5.60,6.20,3.90,4.00, SURF_ID='Wood Joists02'/ Rafter28
&OBST XB=8.60,8.70,6.30,6.90,3.90,4.00, SURF_ID='Wood Joists02'/ Rafter28
&OBST XB=8.60,8.70,6.80,7.30,3.80,3.90, SURF_ID='Wood Joists02'/ Rafter28
&OBST XB=8.60,8.70,7.20,7.70,3.70,3.80, SURF_ID='Wood Joists02'/ Rafter28
&OBST XB=8.60,8.70,7.60,8.10,3.60,3.70, SURF_ID='Wood Joists02'/ Rafter28
&OBST XB=8.60,8.70,8.00,8.50,3.50,3.60, SURF_ID='Wood Joists02'/ Rafter28
&OBST XB=8.60,8.70,8.40,8.90,3.40,3.50, SURF_ID='Wood Joists02'/ Rafter28
&OBST XB=8.60,8.70,8.80,9.30,3.30,3.40, SURF_ID='Wood Joists02'/ Rafter28
&OBST XB=8.60,8.70,9.20,9.70,3.20,3.30, SURF_ID='Wood Joists02'/ Rafter28
&OBST XB=8.60,8.70,9.60,10.10,3.10,3.20, SURF_ID='Wood Joists02'/ Rafter28
&OBST XB=8.60,8.70,10.00,10.50,3.00,3.10, SURF_ID='Wood Joists02'/ Rafter28
&OBST XB=8.60,8.70,10.400,10.90,2.90,3.00, SURF_ID='Wood Joists02'/ Rafter28
&OBST XB=8.60,8.70,10.80,11.30,2.80,2.90, SURF_ID='Wood Joists02'/ Rafter28
&OBST XB=8.60,8.70,11.20,11.70,2.70,2.80, SURF_ID='Wood Joists02'/ Rafter28
&OBST XB=8.00,8.10,0.50,14.90,2.50,2.70, SURF_ID='Wood Joists02'/ Rafter29
&OBST XB=8.00,8.10,0.80,1.30,2.70,2.80, SURF_ID='Wood Joists02'/ Rafter29
&OBST XB=8.00,8.10,1.20,1.70,2.80,2.90, SURF_ID='Wood Joists02'/ Rafter29
&OBST XB=8.00,8.10,1.60,2.10,2.90,3.00, SURF_ID='Wood Joists02'/ Rafter29
&OBST XB=8.00,8.10,2.00,2.50,3.00,3.10, SURF_ID='Wood Joists02'/ Rafter29
&OBST XB=8.00,8.10,2.40,2.90,3.10,3.20, SURF_ID='Wood Joists02'/ Rafter29
&OBST XB=8.00,8.10,2.80,3.30,3.20,3.30, SURF_ID='Wood Joists02'/ Rafter29
&OBST XB=8.00,8.10,3.20,3.70,3.30,3.40, SURF_ID='Wood Joists02'/ Rafter29
&OBST XB=8.00,8.10,3.60,4.10,3.40,3.50, SURF_ID='Wood Joists02'/ Rafter29
&OBST XB=8.00,8.10,4.00,4.50,3.50,3.60, SURF_ID='Wood Joists02'/ Rafter29
&OBST XB=8.00,8.10,4.40,4.90,3.60,3.70, SURF_ID='Wood Joists02'/ Rafter29
&OBST XB=8.00,8.10,4.80,5.30,3.70,3.80, SURF_ID='Wood Joists02'/ Rafter29
&OBST XB=8.00,8.10,5.20,5.70,3.80,3.90, SURF_ID='Wood Joists02'/ Rafter29
&OBST XB=8.00,8.10,5.60,6.20,3.90,4.00, SURF_ID='Wood Joists02'/ Rafter29
&OBST XB=8.00,8.10,6.30,6.90,3.90,4.00, SURF_ID='Wood Joists02'/ Rafter29
&OBST XB=8.00,8.10,6.80,7.30,3.80,3.90, SURF_ID='Wood Joists02'/ Rafter29
&OBST XB=8.00,8.10,7.20,7.70,3.70,3.80, SURF_ID='Wood Joists02'/ Rafter29
&OBST XB=8.00,8.10,7.60,8.10,3.60,3.70, SURF_ID='Wood Joists02'/ Rafter29
&OBST XB=8.00,8.10,8.00,8.50,3.50,3.60, SURF_ID='Wood Joists02'/ Rafter29
&OBST XB=8.00,8.10,8.40,8.90,3.40,3.50, SURF_ID='Wood Joists02'/ Rafter29
&OBST XB=8.00,8.10,8.80,9.30,3.30,3.40, SURF_ID='Wood Joists02'/ Rafter29
&OBST XB=8.00,8.10,9.20,9.70,3.20,3.30, SURF_ID='Wood Joists02'/ Rafter29
&OBST XB=8.00,8.10,9.60,10.10,3.10,3.20, SURF_ID='Wood Joists02'/ Rafter29
&OBST XB=8.00,8.10,10.00,10.50,3.00,3.10, SURF_ID='Wood Joists02'/ Rafter29
&OBST XB=8.00,8.10,10.400,10.90,2.90,3.00, SURF_ID='Wood Joists02'/ Rafter29
&OBST XB=8.00,8.10,10.80,11.30,2.80,2.90, SURF_ID='Wood Joists02'/ Rafter29
&OBST XB=8.00,8.10,11.20,11.70,2.70,2.80, SURF_ID='Wood Joists02'/ Rafter29
&OBST XB=7.40,7.50,0.50,14.90,2.50,2.70, SURF_ID='Wood Joists02'/ Rafter30
&OBST XB=7.40,7.50,0.80,1.30,2.70,2.80, SURF_ID='Wood Joists02'/ Rafter30
&OBST XB=7.40,7.50,1.20,1.70,2.80,2.90, SURF_ID='Wood Joists02'/ Rafter30
```

```
&OBST XB=7.40,7.50,1.60,2.10,2.90,3.00, SURF_ID='Wood Joists02'/ Rafter30
&OBST XB=7.40,7.50,2.00,2.50,3.00,3.10, SURF_ID='Wood Joists02'/ Rafter30
&OBST XB=7.40,7.50,2.40,2.90,3.10,3.20, SURF_ID='Wood Joists02'/ Rafter30
&OBST XB=7.40,7.50,2.80,3.30,3.20,3.30, SURF_ID='Wood Joists02'/ Rafter30
&OBST XB=7.40,7.50,3.60,4.10,3.40,3.50, SURF_ID='Wood Joists02'/ Rafter30
&OBST XB=7.40,7.50,4.00,4.50,3.50,3.60, SURF_ID='Wood Joists02'/ Rafter30
&OBST XB=7.40,7.50,4.40,4.90,3.60,3.70, SURF_ID='Wood Joists02'/ Rafter30
&OBST XB=7.40,7.50,4.80,5.30,3.70,3.80, SURF_ID='Wood Joists02'/ Rafter30
&OBST XB=7.40,7.50,5.20,5.70,3.80,3.90, SURF_ID='Wood Joists02'/ Rafter30
&OBST XB=7.40,7.50,5.60,6.20,3.90,4.00, SURF_ID='Wood Joists02'/ Rafter30
&OBST XB=7.40,7.50,6.30,6.90,3.90,4.00, SURF_ID='Wood Joists02'/ Rafter30
&OBST XB=7.40,7.50,6.80,7.30,3.80,3.90, SURF_ID='Wood Joists02'/ Rafter30
&OBST XB=7.40,7.50,7.20,7.70,3.70,3.80, SURF_ID='Wood Joists02'/ Rafter30
&OBST XB=7.40,7.50,7.60,8.10,3.60,3.70, SURF_ID='Wood Joists02'/ Rafter30
&OBST XB=7.40,7.50,8.00,8.50,3.50,3.60, SURF_ID='Wood Joists02'/ Rafter30
&OBST XB=7.40,7.50,8.40,8.90,3.40,3.50, SURF_ID='Wood Joists02'/ Rafter30
&OBST XB=7.40,7.50,8.80,9.30,3.30,3.40, SURF_ID='Wood Joists02'/ Rafter30
&OBST XB=7.40,7.50,9.20,9.70,3.20,3.30, SURF_ID='Wood Joists02'/ Rafter30
&OBST XB=7.40,7.50,9.60,10.10,3.10,3.20, SURF_ID='Wood Joists02'/ Rafter30
&OBST XB=7.40,7.50,10.00,10.50,3.00,3.10, SURF_ID='Wood Joists02'/ Rafter30
&OBST XB=7.40,7.50,10.400,10.90,2.90,3.00, SURF_ID='Wood Joists02'/ Rafter30
&OBST XB=7.40,7.50,10.80,11.30,2.80,2.90, SURF_ID='Wood Joists02'/ Rafter30
&OBST XB=7.40,7.50,11.20,11.70,2.70,2.80, SURF_ID='Wood Joists02'/ Rafter30
&OBST XB=6.80,6.90,0.50,14.90,2.50,2.70, SURF_ID='Wood Joists02'/ Rafter31
&OBST XB=6.80,6.90,0.80,1.30,2.70,2.80, SURF_ID='Wood Joists02'/ Rafter31
&OBST XB=6.80,6.90,1.20,1.70,2.80,2.90, SURF_ID='Wood Joists02'/ Rafter31
&OBST XB=6.80,6.90,1.60,2.10,2.90,3.00, SURF_ID='Wood Joists02'/ Rafter31
&OBST XB=6.80,6.90,2.00,2.50,3.00,3.10, SURF_ID='Wood Joists02'/ Rafter31
&OBST XB=6.80,6.90,2.40,2.90,3.10,3.20, SURF_ID='Wood Joists02'/ Rafter31
&OBST XB=6.80,6.90,2.80,3.30,3.20,3.30, SURF_ID='Wood Joists02'/ Rafter31
&OBST XB=6.80,6.90,3.20,3.70,3.30,3.40, SURF_ID='Wood Joists02'/ Rafter31
&OBST XB=6.80,6.90,3.60,4.10,3.40,3.50, SURF_ID='Wood Joists02'/ Rafter31
&OBST XB=6.80,6.90,4.00,4.50,3.50,3.60, SURF_ID='Wood Joists02'/ Rafter31
&OBST XB=6.80,6.90,4.40,4.90,3.60,3.70, SURF_ID='Wood Joists02'/ Rafter31
&OBST XB=6.80,6.90,4.80,5.30,3.70,3.80, SURF_ID='Wood Joists02'/ Rafter31
&OBST XB=6.80,6.90,5.20,5.70,3.80,3.90, SURF_ID='Wood Joists02'/ Rafter31
&OBST XB=6.80,6.90,5.60,6.20,3.90,4.00, SURF_ID='Wood Joists02'/ Rafter31
&OBST XB=6.80,6.90,6.30,6.90,3.90,4.00, SURF_ID='Wood Joists02'/ Rafter31
&OBST XB=6.80,6.90,6.80,7.30,3.80,3.90, SURF_ID='Wood Joists02'/ Rafter31
&OBST XB=6.80,6.90,7.20,7.70,3.70,3.80, SURF_ID='Wood Joists02'/ Rafter31
&OBST XB=6.80,6.90,7.60,8.10,3.60,3.70, SURF_ID='Wood Joists02'/ Rafter31
&OBST XB=6.80,6.90,8.00,8.50,3.50,3.60, SURF_ID='Wood Joists02'/ Rafter31
&OBST XB=6.80,6.90,8.40,8.90,3.40,3.50, SURF_ID='Wood Joists02'/ Rafter31
&OBST XB=6.80,6.90,8.80,9.30,3.30,3.40, SURF_ID='Wood Joists02'/ Rafter31
&OBST XB=6.80,6.90,9.20,9.70,3.20,3.30, SURF_ID='Wood Joists02'/ Rafter31
&OBST XB=6.80,6.90,9.60,10.10,3.10,3.20, SURF_ID='Wood Joists02'/ Rafter31
&OBST XB=6.80,6.90,10.00,10.50,3.00,3.10, SURF_ID='Wood Joists02'/ Rafter31
&OBST XB=6.80,6.90,10.400,10.90,2.90,3.00, SURF_ID='Wood Joists02'/ Rafter31
&OBST XB=6.80,6.90,10.80,11.30,2.80,2.90, SURF_ID='Wood Joists02'/ Rafter31
&OBST XB=6.80,6.90,11.20,11.70,2.70,2.80, SURF_ID='Wood Joists02'/ Rafter31
&OBST XB=6.20,6.30,0.50,14.90,2.50,2.70, SURF_ID='Wood Joists02'/ Rafter32
&OBST XB=6.20,6.30,0.80,1.30,2.70,2.80, SURF_ID='Wood Joists02'/ Rafter32
&OBST XB=6.20,6.30,1.20,1.70,2.80,2.90, SURF_ID='Wood Joists02'/ Rafter32
&OBST XB=6.20,6.30,1.60,2.10,2.90,3.00, SURF_ID='Wood Joists02'/ Rafter32
&OBST XB=6.20,6.30,2.00,2.50,3.00,3.10, SURF_ID='Wood Joists02'/ Rafter32
&OBST XB=6.20,6.30,2.40,2.90,3.10,3.20, SURF_ID='Wood Joists02'/ Rafter32
&OBST XB=6.20,6.30,2.80,3.30,3.20,3.30, SURF_ID='Wood Joists02'/ Rafter32
&OBST XB=6.20,6.30,3.20,3.70,3.30,3.40, SURF_ID='Wood Joists02'/ Rafter32
&OBST XB=6.20,6.30,3.60,4.10,3.40,3.50, SURF_ID='Wood Joists02'/ Rafter32
&OBST XB=6.20,6.30,4.00,4.50,3.50,3.60, SURF_ID='Wood Joists02'/ Rafter32
&OBST XB=6.20,6.30,4.40,4.90,3.60,3.70, SURF_ID='Wood Joists02'/ Rafter32
&OBST XB=6.20,6.30,4.80,5.30,3.70,3.80, SURF_ID='Wood Joists02'/ Rafter32
&OBST XB=6.20,6.30,5.20,5.70,3.80,3.90, SURF_ID='Wood Joists02'/ Rafter32
&OBST XB=6.20,6.30,5.60,6.20,3.90,4.00, SURF_ID='Wood Joists02'/ Rafter32
&OBST XB=6.20,6.30,6.30,6.90,3.90,4.00, SURF_ID='Wood Joists02'/ Rafter32
&OBST XB=6.20,6.30,6.80,7.30,3.80,3.90, SURF_ID='Wood Joists02'/ Rafter32
&OBST XB=6.20,6.30,7.20,7.70,3.70,3.80, SURF_ID='Wood Joists02'/ Rafter32
&OBST XB=6.20,6.30,7.60,8.10,3.60,3.70, SURF_ID='Wood Joists02'/ Rafter32
&OBST XB=6.20,6.30,8.00,8.50,3.50,3.60, SURF_ID='Wood Joists02'/ Rafter32
&OBST XB=6.20,6.30,8.40,8.90,3.40,3.50, SURF_ID='Wood Joists02'/ Rafter32
&OBST XB=6.20,6.30,8.80,9.30,3.30,3.40, SURF_ID='Wood Joists02'/ Rafter32
&OBST XB=6.20,6.30,9.20,9.70,3.20,3.30, SURF_ID='Wood Joists02'/ Rafter32
&OBST XB=6.20,6.30,9.60,10.10,3.10,3.20, SURF_ID='Wood Joists02'/ Rafter32
&OBST XB=6.20,6.30,10.00,10.50,3.00,3.10, SURF_ID='Wood Joists02'/ Rafter32
&OBST XB=6.20,6.30,10.400,10.90,2.90,3.00, SURF_ID='Wood Joists02'/ Rafter32
&OBST XB=6.20,6.30,10.80,11.30,2.80,2.90, SURF_ID='Wood Joists02'/ Rafter32
&OBST XB=6.20,6.30,11.20,11.70,2.70,2.80, SURF_ID='Wood Joists02'/ Rafter32
&OBST XB=5.60,5.70,0.50,14.90,2.50,2.70, SURF_ID='Wood Joists02'/ Rafter33
&OBST XB=5.60,5.70,0.80,1.30,2.70,2.80, SURF_ID='Wood Joists02'/ Rafter33
&OBST XB=5.60,5.70,1.20,1.70,2.80,2.90, SURF_ID='Wood Joists02'/ Rafter33
&OBST XB=5.60,5.70,1.60,2.10,2.90,3.00, SURF_ID='Wood Joists02'/ Rafter33
&OBST XB=5.60,5.70,2.00,2.50,3.00,3.10, SURF_ID='Wood Joists02'/ Rafter33
&OBST XB=5.60,5.70,2.40,2.90,3.10,3.20, SURF_ID='Wood Joists02'/ Rafter33
&OBST XB=5.60,5.70,2.80,3.30,3.20,3.30, SURF_ID='Wood Joists02'/ Rafter33
&OBST XB=5.60,5.70,3.20,3.70,3.30,3.40, SURF_ID='Wood Joists02'/ Rafter33
&OBST XB=5.60,5.70,3.60,4.10,3.40,3.50, SURF_ID='Wood Joists02'/ Rafter33
&OBST XB=5.60,5.70,4.00,4.50,3.50,3.60, SURF_ID='Wood Joists02'/ Rafter33
&OBST XB=5.60,5.70,4.40,4.90,3.60,3.70, SURF_ID='Wood Joists02'/ Rafter33
&OBST XB=5.60,5.70,4.80,5.30,3.70,3.80, SURF_ID='Wood Joists02'/ Rafter33
&OBST XB=5.60,5.70,5.20,5.70,3.80,3.90, SURF_ID='Wood Joists02'/ Rafter33
&OBST XB=5.60,5.70,5.60,6.20,3.90,4.00, SURF_ID='Wood Joists02'/ Rafter33
&OBST XB=5.60,5.70,6.30,6.90,3.90,4.00, SURF_ID='Wood Joists02'/ Rafter33
&OBST XB=5.60,5.70,6.80,7.30,3.80,3.90, SURF_ID='Wood Joists02'/ Rafter33
&OBST XB=5.60,5.70,7.20,7.70,3.70,3.80, SURF_ID='Wood Joists02'/ Rafter33
```

```
&OBST XB=5.60,5.70,7.60,8.10,3.60,3.70, SURF_ID='Wood Joists02'/ Rafter33
&OBST XB=5.60,5.70,8.00,8.50,3.50,3.60, SURF_ID='Wood Joists02'/ Rafter33
&OBST XB=5.60,5.70,8.40,8.90,3.40,3.50, SURF_ID='Wood Joists02'/ Rafter33
&OBST XB=5.60,5.70,8.80,9.30,3.30,3.40, SURF_ID='Wood Joists02'/ Rafter33
&OBST XB=5.60,5.70,9.20,9.70,3.20,3.30, SURF_ID='Wood Joists02'/ Rafter33
&OBST XB=5.60,5.70,9.60,10.10,3.10,3.20, SURF_ID='Wood Joists02'/ Rafter33
&OBST XB=5.60,5.70,10.00,10.50,3.00,3.10, SURF_ID='Wood Joists02'/ Rafter33
&OBST XB=5.60,5.70,10.400,10.90,2.90,3.00, SURF_ID='Wood Joists02'/ Rafter33
&OBST XB=5.60,5.70,10.80,11.30,2.80,2.90, SURF_ID='Wood Joists02'/ Rafter33
&OBST XB=5.60,5.70,11.20,11.70,2.70,2.80, SURF_ID='Wood Joists02'/ Rafter33
&OBST XB=5.00,5.10,0.50,14.90,2.50,2.70, SURF_ID='Wood Joists02'/ Rafter34
&OBST XB=5.00,5.10,0.80,1.30,2.70,2.80, SURF_ID='Wood Joists02'/ Rafter34
&OBST XB=5.00,5.10,1.20,1.70,2.80,2.90, SURF_ID='Wood Joists02'/ Rafter34
&OBST XB=5.00,5.10,1.60,2.10,2.90,3.00, SURF_ID='Wood Joists02'/ Rafter34
&OBST XB=5.00,5.10,2.00,2.50,3.00,3.10, SURF_ID='Wood Joists02'/ Rafter34
&OBST XB=5.00,5.10,2.40,2.90,3.10,3.20, SURF_ID='Wood Joists02'/ Rafter34
&OBST XB=5.00,5.10,2.80,3.30,3.20,3.30, SURF_ID='Wood Joists02'/ Rafter34
&OBST XB=5.00,5.10,3.20,3.70,3.30,3.40, SURF_ID='Wood Joists02'/ Rafter34
&OBST XB=5.00,5.10,3.60,4.10,3.40,3.50, SURF_ID='Wood Joists02'/ Rafter34
&OBST XB=5.00,5.10,4.00,4.50,3.50,3.60, SURF_ID='Wood Joists02'/ Rafter34
&OBST XB=5.00,5.10,4.40,4.90,3.60,3.70, SURF_ID='Wood Joists02'/ Rafter34
&OBST XB=5.00,5.10,4.80,5.30,3.70,3.80, SURF_ID='Wood Joists02'/ Rafter34
&OBST XB=5.00,5.10,5.20,5.70,3.80,3.90, SURF_ID='Wood Joists02'/ Rafter34
&OBST XB=5.00,5.10,5.60,6.20,3.90,4.00, SURF_ID='Wood Joists02'/ Rafter34
&OBST XB=5.00,5.10,6.30,6.90,3.90,4.00, SURF_ID='Wood Joists02'/ Rafter34
&OBST XB=5.00,5.10,6.80,7.30,3.80,3.90, SURF_ID='Wood Joists02'/ Rafter34
&OBST XB=5.00,5.10,7.20,7.70,3.70,3.80, SURF_ID='Wood Joists02'/ Rafter34
&OBST XB=5.00,5.10,7.60,8.10,3.60,3.70, SURF_ID='Wood Joists02'/ Rafter34
&OBST XB=5.00,5.10,8.00,8.50,3.50,3.60, SURF_ID='Wood Joists02'/ Rafter34
&OBST XB=5.00,5.10,8.40,8.90,3.40,3.50, SURF_ID='Wood Joists02'/ Rafter34
&OBST XB=5.00,5.10,8.80,9.30,3.30,3.40, SURF_ID='Wood Joists02'/ Rafter34
&OBST XB=5.00,5.10,9.20,9.70,3.20,3.30, SURF_ID='Wood Joists02'/ Rafter34
&OBST XB=5.00,5.10,9.60,10.10,3.10,3.20, SURF_ID='Wood Joists02'/ Rafter34
&OBST XB=5.00,5.10,10.00,10.50,3.00,3.10, SURF_ID='Wood Joists02'/ Rafter34
&OBST XB=5.00,5.10,10.400,10.90,2.90,3.00, SURF_ID='Wood Joists02'/ Rafter34
&OBST XB=5.00,5.10,10.80,11.30,2.80,2.90, SURF_ID='Wood Joists02'/ Rafter34
&OBST XB=5.00,5.10,11.20,11.70,2.70,2.80, SURF_ID='Wood Joists02'/ Rafter34
&OBST XB=4.40,4.50,0.50,14.90,2.50,2.70, SURF_ID='Wood Joists02'/ Rafter35
&OBST XB=4.40,4.50,0.80,1.30,2.70,2.80, SURF_ID='Wood Joists02'/ Rafter35
&OBST XB=4.40,4.50,1.20,1.70,2.80,2.90, SURF_ID='Wood Joists02'/ Rafter35
&OBST XB=4.40,4.50,1.60,2.10,2.90,3.00, SURF_ID='Wood Joists02'/ Rafter35
&OBST XB=4.40,4.50,2.00,2.50,3.00,3.10, SURF_ID='Wood Joists02'/ Rafter35
&OBST XB=4.40,4.50,2.40,2.90,3.10,3.20, SURF_ID='Wood Joists02'/ Rafter35
&OBST XB=4.40,4.50,2.80,3.30,3.20,3.30, SURF_ID='Wood Joists02'/ Rafter35
&OBST XB=4.40,4.50,3.20,3.70,3.30,3.40, SURF_ID='Wood Joists02'/ Rafter35
&OBST XB=4.40,4.50,3.60,4.10,3.40,3.50, SURF_ID='Wood Joists02'/ Rafter35
&OBST XB=4.40,4.50,4.00,4.50,3.50,3.60, SURF_ID='Wood Joists02'/ Rafter35
&OBST XB=4.40,4.50,4.40,4.90,3.60,3.70, SURF_ID='Wood Joists02'/ Rafter35
&OBST XB=4.40,4.50,4.80,5.30,3.70,3.80, SURF_ID='Wood Joists02'/ Rafter35
&OBST XB=4.40,4.50,5.20,5.70,3.80,3.90, SURF_ID='Wood Joists02'/ Rafter35
&OBST XB=4.40,4.50,5.60,6.20,3.90,4.00, SURF_ID='Wood Joists02'/ Rafter35
&OBST XB=4.40,4.50,6.30,6.90,3.90,4.00, SURF_ID='Wood Joists02'/ Rafter35
&OBST XB=4.40,4.50,6.80,7.30,3.80,3.90, SURF_ID='Wood Joists02'/ Rafter35
&OBST XB=4.40,4.50,7.20,7.70,3.70,3.80, SURF_ID='Wood Joists02'/ Rafter35
&OBST XB=4.40,4.50,7.60,8.10,3.60,3.70, SURF_ID='Wood Joists02'/ Rafter35
&OBST XB=4.40,4.50,8.00,8.50,3.50,3.60, SURF_ID='Wood Joists02'/ Rafter35
&OBST XB=4.40,4.50,8.40,8.90,3.40,3.50, SURF_ID='Wood Joists02'/ Rafter35
&OBST XB=4.40,4.50,8.80,9.30,3.30,3.40, SURF_ID='Wood Joists02'/ Rafter35
&OBST XB=4.40,4.50,9.20,9.70,3.20,3.30, SURF_ID='Wood Joists02'/ Rafter35
&OBST XB=4.40,4.50,9.60,10.10,3.10,3.20, SURF_ID='Wood Joists02'/ Rafter35
&OBST XB=4.40,4.50,10.00,10.50,3.00,3.10, SURF_ID='Wood Joists02'/ Rafter35
&OBST XB=4.40,4.50,10.400,10.90,2.90,3.00, SURF_ID='Wood Joists02'/ Rafter35
&OBST XB=4.40,4.50,10.80,11.30,2.80,2.90, SURF_ID='Wood Joists02'/ Rafter35
&OBST XB=4.40,4.50,11.20,11.70,2.70,2.80, SURF_ID='Wood Joists02'/ Rafter35
&OBST XB=3.80,3.90,0.50,14.90,2.50,2.70, SURF_ID='Wood Joists02'/ Rafter36
&OBST XB=3.80,3.90,0.80,1.30,2.70,2.80, SURF_ID='Wood Joists02'/ Rafter36
&OBST XB=3.80,3.90,1.20,1.70,2.80,2.90, SURF_ID='Wood Joists02'/ Rafter36
&OBST XB=3.80,3.90,1.60,2.10,2.90,3.00, SURF_ID='Wood Joists02'/ Rafter36
&OBST XB=3.80,3.90,2.00,2.50,3.00,3.10, SURF_ID='Wood Joists02'/ Rafter36
&OBST XB=3.80,3.90,2.40,2.90,3.10,3.20, SURF_ID='Wood Joists02'/ Rafter36
&OBST XB=3.80,3.90,2.80,3.30,3.20,3.30, SURF_ID='Wood Joists02'/ Rafter36
&OBST XB=3.80,3.90,3.20,3.70,3.30,3.40, SURF_ID='Wood Joists02'/ Rafter36
&OBST XB=3.80,3.90,3.60,4.10,3.40,3.50, SURF_ID='Wood Joists02'/ Rafter36
&OBST XB=3.80,3.90,4.00,4.50,3.50,3.60, SURF_ID='Wood Joists02'/ Rafter36
&OBST XB=3.80,3.90,4.40,4.90,3.60,3.70, SURF_ID='Wood Joists02'/ Rafter36
&OBST XB=3.80,3.90,4.80,5.30,3.70,3.80, SURF_ID='Wood Joists02'/ Rafter36
&OBST XB=3.80,3.90,5.20,7.20,3.80,3.90, SURF_ID='Wood Joists02'/ Rafter36
&OBST XB=3.80,3.90,7.20,7.70,3.70,3.80, SURF_ID='Wood Joists02'/ Rafter36
&OBST XB=3.80,3.90,7.60,8.10,3.60,3.70, SURF_ID='Wood Joists02'/ Rafter36
&OBST XB=3.80,3.90,8.00,8.50,3.50,3.60, SURF_ID='Wood Joists02'/ Rafter36
&OBST XB=3.80,3.90,8.40,8.90,3.40,3.50, SURF_ID='Wood Joists02'/ Rafter36
&OBST XB=3.80,3.90,8.80,9.30,3.30,3.40, SURF_ID='Wood Joists02'/ Rafter36
&OBST XB=3.80,3.90,9.20,9.70,3.20,3.30, SURF_ID='Wood Joists02'/ Rafter36
&OBST XB=3.80,3.90,9.60,10.10,3.10,3.20, SURF_ID='Wood Joists02'/ Rafter36
&OBST XB=3.80,3.90,10.00,10.50,3.00,3.10, SURF_ID='Wood Joists02'/ Rafter36
&OBST XB=3.80,3.90,10.400,10.90,2.90,3.00, SURF_ID='Wood Joists02'/ Rafter36
&OBST XB=3.80,3.90,10.80,11.30,2.80,2.90, SURF_ID='Wood Joists02'/ Rafter36
&OBST XB=3.80,3.90,11.20,11.70,2.70,2.80, SURF_ID='Wood Joists02'/ Rafter36
&OBST XB=3.40,3.50,0.90,1.30,2.70,2.80, SURF_ID='Wood Joists02'/ Rafter37
&OBST XB=3.40,3.50,1.30,1.70,2.80,2.90, SURF_ID='Wood Joists02'/ Rafter37
&OBST XB=3.40,3.50,1.70,2.10,2.90,3.00, SURF_ID='Wood Joists02'/ Rafter37
&OBST XB=3.40,3.50,2.10,2.50,3.00,3.10, SURF_ID='Wood Joists02'/ Rafter37
&OBST XB=3.40,3.50,2.50,2.90,3.10,3.20, SURF_ID='Wood Joists02'/ Rafter37
&OBST XB=3.40,3.50,2.90,3.30,3.20,3.30, SURF_ID='Wood Joists02'/ Rafter37
&OBST XB=3.40,3.50,3.30,3.70,3.30,3.40, SURF_ID='Wood Joists02'/ Rafter37
```

```
&OBST XB=3.40,3.50,3.70,4.10,3.40,3.50, SURF_ID='Wood Joists02'/ Rafter37
&OBST XB=3.40,3.50,4.10,4.50,3.50,3.60, SURF_ID='Wood Joists02'/ Rafter37
&OBST XB=3.40,3.50,4.50,4.90,3.60,3.70, SURF_ID='Wood Joists02'/ Rafter37
&OBST XB=3.40,3.50,4.90,7.60,3.70,3.80, SURF_ID='Wood Joists02'/ Rafter37
&OBST XB=3.40,3.50,0.50,14.90,2.50,2.70, SURF_ID='Wood Joists02'/ Rafter37
&OBST XB=3.40,3.50,11.20,11.60,2.70,2.80, SURF_ID='Wood Joists02'/ Rafter37
&OBST XB=3.40,3.50,10.80,11.20,2.80,2.90, SURF_ID='Wood Joists02'/ Rafter37
&OBST XB=3.40,3.50,10.400,10.80,2.90,3.00, SURF_ID='Wood Joists02'/ Rafter37
&OBST XB=3.40,3.50,10.00,10.400,3.00,3.10, SURF_ID='Wood Joists02'/ Rafter37
&OBST XB=3.40,3.50,9.60,10.00,3.10,3.20, SURF_ID='Wood Joists02'/ Rafter37
&OBST XB=3.40,3.50,9.20,9.60,3.20,3.30, SURF_ID='Wood Joists02'/ Rafter37
&OBST XB=3.40,3.50,8.80,9.20,3.30,3.40, SURF_ID='Wood Joists02'/ Rafter37
&OBST XB=3.40,3.50,8.40,8.80,3.40,3.50, SURF_ID='Wood Joists02'/ Rafter37
&OBST XB=3.40,3.50,8.00,8.40,3.50,3.60, SURF_ID='Wood Joists02'/ Rafter37
&OBST XB=3.40,3.50,7.60,8.00,3.60,3.70, SURF_ID='Wood Joists02'/ Rafter37
&OBST XB=2.80,2.90,0.90,1.30,2.70,2.80, SURF_ID='Wood Joists02'/ Rafter38
&OBST XB=2.80,2.90,1.30,1.70,2.80,2.90, SURF_ID='Wood Joists02'/ Rafter38
&OBST XB=2.80,2.90,1.70,2.10,2.90,3.00, SURF_ID='Wood Joists02'/ Rafter38
&OBST XB=2.80,2.90,2.10,2.50,3.00,3.10, SURF_ID='Wood Joists02'/ Rafter38
&OBST XB=2.80,2.90,2.50,2.90,3.10,3.20, SURF_ID='Wood Joists02'/ Rafter38
&OBST XB=2.80,2.90,2.90,3.30,3.20,3.30, SURF_ID='Wood Joists02'/ Rafter38
&OBST XB=2.80,2.90,3.30,3.70,3.30,3.40, SURF_ID='Wood Joists02'/ Rafter38
&OBST XB=2.80,2.90,3.70,4.10,3.40,3.50, SURF_ID='Wood Joists02'/ Rafter38
&OBST XB=2.80,2.90,4.10,4.50,3.50,3.60, SURF_ID='Wood Joists02'/ Rafter38
&OBST XB=2.80,2.90,0.50,14.90,2.50,2.70, SURF_ID='Wood Joists02'/ Rafter38
&OBST XB=2.80,2.90,11.20,11.60,2.70,2.80, SURF_ID='Wood Joists02'/ Rafter38
&OBST XB=2.80,2.90,10.80,11.20,2.80,2.90, SURF_ID='Wood Joists02'/ Rafter38
&OBST XB=2.80,2.90,10.400,10.80,2.90,3.00, SURF_ID='Wood Joists02'/ Rafter38
&OBST XB=2.80,2.90,10.00,10.400,3.00,3.10, SURF_ID='Wood Joists02'/ Rafter38
&OBST XB=2.80,2.90,9.60,10.00,3.10,3.20, SURF_ID='Wood Joists02'/ Rafter38
&OBST XB=2.80,2.90,9.20,9.60,3.20,3.30, SURF_ID='Wood Joists02'/ Rafter38
&OBST XB=2.80,2.90,8.80,9.20,3.30,3.40, SURF_ID='Wood Joists02'/ Rafter38
&OBST XB=2.80,2.90,8.40,8.80,3.40,3.50, SURF_ID='Wood Joists02'/ Rafter38
&OBST XB=2.80,2.90,8.00,8.40,3.50,3.60, SURF_ID='Wood Joists02'/ Rafter38
&OBST XB=2.80,2.90,7.60,8.00,3.60,3.70, SURF_ID='Wood Joists02'/ Rafter38
&OBST XB=2.20,2.30,0.90,1.30,2.70,2.80, SURF_ID='Wood Joists02'/ Rafter39
&OBST XB=2.20,2.30,1.30,1.70,2.80,2.90, SURF_ID='Wood Joists02'/ Rafter39
&OBST XB=2.20,2.30,1.70,2.10,2.90,3.00, SURF_ID='Wood Joists02'/ Rafter39
&OBST XB=2.20,2.30,2.10,2.50,3.00,3.10, SURF_ID='Wood Joists02'/ Rafter39
&OBST XB=2.20,2.30,2.50,2.90,3.10,3.20, SURF_ID='Wood Joists02'/ Rafter39
&OBST XB=2.20,2.30,2.90,3.30,3.20,3.30, SURF_ID='Wood Joists02'/ Rafter39
&OBST XB=2.20,2.30,3.30,3.70,3.30,3.40, SURF_ID='Wood Joists02'/ Rafter39
&OBST XB=2.20,2.30,3.70,3.90,3.40,3.50, SURF_ID='Wood Joists02'/ Rafter39
&OBST XB=2.20,2.30,0.50,12.00,2.50,2.70, SURF_ID='Wood Joists02'/ Rafter39
&OBST XB=2.20,2.30,11.20,11.60,2.70,2.80, SURF_ID='Wood Joists02'/ Rafter39
&OBST XB=2.20,2.30,10.80,11.20,2.80,2.90, SURF_ID='Wood Joists02'/ Rafter39
&OBST XB=2.20,2.30,10.400,10.80,2.90,3.00, SURF_ID='Wood Joists02'/ Rafter39
&OBST XB=2.20,2.30,10.00,10.400,3.00,3.10, SURF_ID='Wood Joists02'/ Rafter39
&OBST XB=2.20,2.30,9.60,10.00,3.10,3.20, SURF_ID='Wood Joists02'/ Rafter39
&OBST XB=2.20,2.30,9.20,9.60,3.20,3.30, SURF_ID='Wood Joists02'/ Rafter39
&OBST XB=2.20,2.30,8.80,9.20,3.30,3.40, SURF_ID='Wood Joists02'/ Rafter39
&OBST XB=2.20,2.30,8.20,8.80,3.40,3.50, SURF_ID='Wood Joists02'/ Rafter39
&OBST XB=1.60,1.70,0.90,1.30,2.70,2.80, SURF_ID='Wood Joists02'/ Rafter40
&OBST XB=1.60,1.70,1.30,1.70,2.80,2.90, SURF_ID='Wood Joists02'/ Rafter40
&OBST XB=1.60,1.70,1.70,2.10,2.90,3.00, SURF_ID='Wood Joists02'/ Rafter40
&OBST XB=1.60,1.70,2.10,2.50,3.00,3.10, SURF_ID='Wood Joists02'/ Rafter40
&OBST XB=1.60,1.70,2.50,2.90,3.10,3.20, SURF_ID='Wood Joists02'/ Rafter40
&OBST XB=1.60,1.70,2.90,3.30,3.20,3.30, SURF_ID='Wood Joists02'/ Rafter40
&OBST XB=1.60,1.70,0.50,12.00,2.50,2.70, SURF_ID='Wood Joists02'/ Rafter40
&OBST XB=1.60,1.70,11.20,11.60,2.70,2.80, SURF_ID='Wood Joists02'/ Rafter40
&OBST XB=1.60,1.70,10.80,11.20,2.80,2.90, SURF_ID='Wood Joists02'/ Rafter40
&OBST XB=1.60,1.70,10.400,10.80,2.90,3.00, SURF_ID='Wood Joists02'/ Rafter40
&OBST XB=1.60,1.70,10.00,10.400,3.00,3.10, SURF_ID='Wood Joists02'/ Rafter40
&OBST XB=1.60,1.70,9.60,10.00,3.10,3.20, SURF_ID='Wood Joists02'/ Rafter40
&OBST XB=1.60,1.70,9.20,9.60,3.20,3.30, SURF_ID='Wood Joists02'/ Rafter40
&OBST XB=1.60,1.70,8.80,9.20,3.30,3.40, SURF_ID='Wood Joists02'/ Rafter40
&OBST XB=1.00,1.10,0.90,1.30,2.70,2.80, SURF_ID='Wood Joists02'/ Rafter41
&OBST XB=1.00,1.10,1.30,1.70,2.80,2.90, SURF_ID='Wood Joists02'/ Rafter41
&OBST XB=1.00,1.10,1.70,2.10,2.90,3.00, SURF_ID='Wood Joists02'/ Rafter41
&OBST XB=1.00,1.10,2.50,2.70,3.10,3.20, SURF_ID='Wood Joists02'/ Rafter41
&OBST XB=1.00,1.10,0.50,12.00,2.50,2.70, SURF_ID='Wood Joists02'/ Rafter41
&OBST XB=1.00,1.10,11.20,11.60,2.70,2.80, SURF_ID='Wood Joists02'/ Rafter41
&OBST XB=1.00,1.10,10.80,11.20,2.80,2.90, SURF_ID='Wood Joists02'/ Rafter41
&OBST XB=1.00,1.10,10.400,10.80,2.90,3.00, SURF_ID='Wood Joists02'/ Rafter41
&OBST XB=1.00,1.10,10.00,10.400,3.00,3.10, SURF_ID='Wood Joists02'/ Rafter41
&OBST XB=1.00,1.10,9.40,10.00,3.10,3.20, SURF_ID='Wood Joists02'/ Rafter41
&OBST XB=0.40,0.50,0.90,1.30,2.70,2.80, SURF_ID='Wood Joists02'/ Rafter42
&OBST XB=0.40,0.50,1.30,1.70,2.80,2.90, SURF_ID='Wood Joists02'/ Rafter42
&OBST XB=0.40,0.50,1.70,2.10,2.90,3.00, SURF_ID='Wood Joists02'/ Rafter42
&OBST XB=0.40,0.50,0.50,12.00,2.50,2.70, SURF_ID='Wood Joists02'/ Rafter42
&OBST XB=0.40,0.50,11.20,11.60,2.70,2.80, SURF_ID='Wood Joists02'/ Rafter42
&OBST XB=0.40,0.50,10.80,11.20,2.80,2.90, SURF_ID='Wood Joists02'/ Rafter42
&OBST XB=0.40,0.50,10.400,10.80,2.90,3.00, SURF_ID='Wood Joists02'/ Rafter42
&OBST XB=0.40,0.50,10.00,10.400,3.00,3.10, SURF_ID='Wood Joists02'/ Rafter42
&OBST XB=-0.2000,-0.1,0.90,1.30,2.70,2.80, SURF_ID='Wood Joists02'/ Rafter43
&OBST XB=-0.2000,-0.1,0.50,12.00,2.50,2.70, SURF_ID='Wood Joists02'/ Rafter43
&OBST XB=-0.2000,-0.1,11.20,11.60,2.70,2.80, SURF_ID='Wood Joists02'/ Rafter43
&OBST XB=-0.2000,-0.1,10.80,11.20,2.80,2.90, SURF_ID='Wood Joists02'/ Rafter43
&OBST XB=11.70,11.80,8.90,9.90,2.50,3.50, SURF_ID6='Roof Sheathing','Gypsum','Roof Sheathing','Roof Sheathing','Roof
Sheathing','Roof Sheathing'/ Skylight Assembly
&OBST XB=11.70,12.90,8.80,8.89,2.50,3.60, SURF_ID6='Roof Sheathing','Roof Sheathing','Roof Sheathing','Gypsum','Roof
Sheathing','Roof Sheathing'/ Skylight Assembly
&OBST XB=12.80,12.90,8.90,9.90,2.50,3.39, SURF_ID6='Gypsum','Roof Sheathing','Roof Sheathing','Roof Sheathing','Roof
Sheathing','Roof Sheathing'/ Skylight Assembly
```

```
&OBST XB=11.70,12.90,9.90,10.00,2.50,3.39, SURF_ID6='Roof Sheathing','Roof Sheathing','Gypsum','Roof Sheathing','Roof
Sheathing','Roof Sheathing'/ Skylight Assembly
&OBST XB=11.80,12.80,8.90,9.90,3.40,3.50, COLOR='BLACK', PERMIT_HOLE=.FALSE., SURF_ID='Overhang Matl'/ Skylight Assembly
&OBST XB=23.00,23.70,8.90,10.20,3.60,3.70, PERMIT_HOLE=.FALSE., SURF_ID='Glass'/ Skylight Glass
&OBST XB=23.00,23.70,10.20,10.30,2.50,3.60, PERMIT_HOLE=.FALSE., SURF_ID='Gypsum'/ Skylight1
&OBST XB=23.70,23.80,9.00,10.30,2.50,3.70, PERMIT_HOLE=.FALSE., SURF_ID='Gypsum'/ Skylight2
&OBST XB=22.90,23.00,9.00,10.30,2.50,3.50, PERMIT_HOLE=.FALSE., SURF_ID='Gypsum'/ Skylight3
&OBST XB=22.90,23.80,8.90,9.00,2.50,3.60, PERMIT_HOLE=.FALSE., SURF_ID='Gypsum'/ Skylight4
&OBST XB=24.90,30.80,-0.60,0.40,2.40,2.50, RGB=153,102,0, SURF_ID='Overhang Matl'/ Roof_Overhang A1
&OBST XB=21.50,24.90,2.10,3.40,2.40,2.50, RGB=153,102,0, SURF_ID='Overhang Matl'/ Roof_Overhang A2
&OBST XB=20.20,24.90,0.40,2.10,2.40,2.50, RGB=153,102,0, SURF_ID='Overhang Matl'/ Roof_Overhang A3
&OBST XB=2.60,20.20,0.40,2.10,2.40,2.50, RGB=153,102,0, SURF_ID='Overhang Matl'/ Roof_Overhang A4
&OBST XB=0.2000,2.60,0.40,2.10,2.40,2.50, RGB=153,102,0, SURF_ID='Overhang Matl'/ Roof_Overhang A5
&OBST XB=-1.30,0.2000,0.40,12.10,2.40,2.50, RGB=153,102,0, SURF_ID='Overhang Matl'/ Roof_Overhang B1
&OBST XB=20.20,30.80,17.40,18.50,2.40,2.50, RGB=153,102,0, SURF_ID='Overhang Matl'/ Roof_Overhang
&OBST XB=2.60,15.30,15.00,16.50,2.40,2.50, RGB=153,102,0, SURF_ID='Overhang Matl'/ Roof_Overhang
&OBST XB=23.00,24.60,5.90,6.00,4.00,4.10, PERMIT_HOLE=.FALSE., SURF_IDS='Overhang Matl','Roof Sheathing','Roof Sheathing02'/ Roof
Sheathing
&OBST XB=24.60,24.80,6.00,6.50,4.10,4.20, PERMIT_HOLE=.FALSE., SURF_IDS='Overhang Matl','Roof Sheathing','Roof Sheathing'/ Roof
Sheathing
&OBST XB=24.60,24.80,5.60,6.00,4.00,4.10, PERMIT_HOLE=.FALSE., SURF_IDS='Overhang Matl','Roof Sheathing','Roof Sheathing'/ Roof
Sheathing
&OBST XB=23.00,24.60,5.60,5.90,4.00,4.10, PERMIT_HOLE=.FALSE., SURF_IDS='Overhang Matl','Roof Sheathing','Roof Sheathing02'/ Roof
Sheathing
&OBST XB=24.60,24.80,6.90,7.30,3.90,4.00, PERMIT_HOLE=.FALSE., SURF_ID='Roof Sheathing'/ Roof Sheathing
&OBST XB=24.60,24.80,6.50,6.90,4.00,4.10, PERMIT_HOLE=.FALSE., SURF_IDS='Overhang Matl','Roof Sheathing','Roof Sheathing'/ Roof
Sheathing
&OBST XB=23.00,24.60,7.30,7.70,3.80,3.90, PERMIT_HOLE=.FALSE., SURF_IDS='Overhang Matl','Roof Sheathing','Roof Sheathing'/ Roof
Sheathing
&OBST XB=23.00,24.60,6.90,7.30,3.90,4.00, PERMIT_HOLE=.FALSE., SURF_IDS='Overhang Matl','Roof Sheathing','Roof Sheathing'/ Roof
Sheathing
&OBST XB=23.00,24.60,6.50,6.90,4.00,4.10, PERMIT_HOLE=.FALSE., SURF_IDS='Overhang Matl','Roof Sheathing','Roof Sheathing02'/ Roof
Sheathing
&OBST XB=14.80,29.90,12.80,13.70,2.70,2.80, SURF_IDS='Overhang Matl','Roof Sheathing','Roof Sheathing'/ Roof_Sheathing
&OBST XB=13.80,29.50,10.90,11.80,2.90,3.00, SURF_IDS='Overhang Matl','Roof Sheathing','Roof Sheathing'/ Roof_Sheathing
&OBST XB=-0.80,2.60,11.70,12.10,2.70,2.80, SURF_IDS='Overhang Matl','Roof Sheathing','Roof Sheathing'/ Roof_Sheathing
&OBST XB=-0.90,3.10,11.60,12.00,2.80,2.90, SURF_IDS='Overhang Matl','Roof Sheathing','Roof Sheathing'/ Roof_Sheathing
&OBST XB=2.60,3.10,11.90,12.10,2.80,2.90, SURF_IDS='Overhang Matl','Roof Sheathing','Roof Sheathing'/ Roof_Sheathing
&OBST XB=19.90,21.30,5.20,5.40,3.90,4.00, PERMIT_HOLE=.FALSE., SURF_IDS='Overhang Matl','Roof Sheathing','Roof Sheathing',
DEVC_ID='TIMER'/ Roof Hole Fix
&OBST XB=19.90,21.30,4.80,5.20,3.80,3.90, PERMIT_HOLE=.FALSE., SURF_IDS='Overhang Matl','Roof Sheathing','Roof Sheathing',
DEVC_ID='TIMER'/ Roof Hole Fix
&OBST XB=19.90,21.30,4.40,4.80,3.70,3.80, PERMIT_HOLE=.FALSE., SURF_IDS='Overhang Matl','Roof Sheathing','Roof Sheathing',
DEVC_ID='TIMER'/ Roof Hole Fix
&OBST XB=19.90,21.30,4.20,4.40,3.60,3.70, PERMIT_HOLE=.FALSE., SURF_IDS='Overhang Matl','Roof Sheathing','Roof Sheathing',
DEVC_ID='TIMER'/ Roof Hole Fix
&OBST XB=-0.1,24.90,1.60,2.10,3.00,3.10, SURF_IDS='Overhang Matl','Roof Sheathing','Roof Sheathing'/ Roof_Sheathing_Main
&OBST XB=1.10,25.00,2.80,3.30,3.30,3.40, SURF_IDS='Overhang Matl','Roof Sheathing','Roof Sheathing'/ Roof_Sheathing_Main
&OBST XB=1.50,25.00,3.20,3.70,3.40,3.50, SURF_IDS='Overhang Matl','Roof Sheathing','Roof Sheathing'/ Roof_Sheathing_Main
&OBST XB=1.90,24.90,3.60,4.10,3.50,3.60, SURF_IDS='Overhang Matl','Roof Sheathing','Roof Sheathing'/ Roof_Sheathing_Main
&OBST XB=2.30,24.90,4.00,4.50,3.60,3.70, SURF_IDS='Overhang Matl','Roof Sheathing','Roof Sheathing'/ Roof_Sheathing_Main
&OBST XB=2.70,24.90,4.40,4.90,3.70,3.80, SURF_IDS='Overhang Matl','Roof Sheathing','Roof Sheathing'/ Roof_Sheathing_Main
&OBST XB=3.10,24.90,4.80,5.30,3.80,3.90, SURF_IDS='Overhang Matl','Roof Sheathing','Roof Sheathing'/ Roof_Sheathing_Main
&OBST XB=-0.90,24.90,0.80,1.30,2.80,2.90, SURF_IDS='Overhang Matl','Roof Sheathing','Roof Sheathing'/ Roof_Sheathing_Main
&OBST XB=-0.50,24.90,1.20,1.70,2.90,3.00, SURF_IDS='Overhang Matl','Roof Sheathing','Roof Sheathing'/ Roof_Sheathing_Main
&OBST XB=0.3,24.90,2.00,2.50,3.10,3.20, SURF_IDS='Overhang Matl','Roof Sheathing','Roof Sheathing'/ Roof_Sheathing_Main
&OBST XB=0.70,24.90,2.40,2.90,3.20,3.30, SURF_IDS='Overhang Matl','Roof Sheathing','Roof Sheathing'/ Roof_Sheathing_Main
&OBST XB=3.10,23.00,7.20,7.70,3.80,3.90, SURF_IDS='Overhang Matl','Roof Sheathing','Roof Sheathing'/ Roof_Sheathing_Main
&OBST XB=3.50,23.00,6.90,7.30,3.90,4.00, SURF_IDS='Overhang Matl','Roof Sheathing','Roof Sheathing'/ Roof_Sheathing_Main
&OBST XB=14.20,29.90,11.80,12.80,2.80,2.90, SURF_IDS='Overhang Matl','Roof Sheathing','Roof Sheathing'/ Roof_Sheathing_Main
&OBST XB=-0.50,4.10,10.90,11.60,2.90,3.00, SURF_IDS='Overhang Matl','Roof Sheathing','Roof Sheathing'/ Roof_Sheathing_Main
&OBST XB=-0.1,29.10,10.400,11.00,3.00,3.10, SURF_IDS='Overhang Matl','Roof Sheathing','Roof Sheathing'/ Roof_Sheathing_Main
&OBST XB=0.3,28.70,10.00,10.50,3.10,3.20, SURF_IDS='Overhang Matl','Roof Sheathing','Roof Sheathing'/ Roof_Sheathing_Main
&OBST XB=0.70,28.50,9.60,10.10,3.20,3.30, SURF_IDS='Overhang Matl','Roof Sheathing','Roof Sheathing'/ Roof_Sheathing_Main
&OBST XB=1.10,27.90,9.20,9.70,3.30,3.40, SURF_IDS='Overhang Matl','Roof Sheathing','Roof Sheathing'/ Roof_Sheathing_Main
&OBST XB=1.50,27.50,8.80,9.30,3.40,3.50, SURF_IDS='Overhang Matl','Roof Sheathing','Roof Sheathing'/ Roof_Sheathing_Main
&OBST XB=1.90,23.30,8.40,8.90,3.50,3.60, SURF_IDS='Overhang Matl','Roof Sheathing','Roof Sheathing'/ Roof_Sheathing_Main
&OBST XB=2.30,23.70,8.00,8.50,3.60,3.70, SURF_IDS='Overhang Matl','Roof Sheathing','Roof Sheathing'/ Roof_Sheathing_Main
&OBST XB=2.70,24.00,7.60,8.10,3.70,3.80, SURF_IDS='Overhang Matl','Roof Sheathing','Roof Sheathing'/ Roof_Sheathing_Main
&OBST XB=3.90,23.00,6.40,6.90,4.00,4.10, SURF_IDS='Overhang Matl','Roof Sheathing','Roof Sheathing'/ Roof_Sheathing_Main
&OBST XB=-1.30,24.90,0.40,0.90,2.70,2.80, SURF_IDS='Overhang Matl','Roof Sheathing','Roof Sheathing'/ Roof_Sheathing_Main
&OBST XB=3.90,23.00,5.60,6.10,4.00,4.10, SURF_IDS='Overhang Matl','Roof Sheathing','Roof Sheathing'/ Roof_Sheathing_Main
&OBST XB=3.50,24.80,5.20,5.70,3.90,4.00, SURF_IDS='Overhang Matl','Roof Sheathing','Roof Sheathing'/ Roof_Sheathing_Main
&OBST XB=9.70,16.00,6.00,6.50,4.10,4.20, SURF_IDS='Overhang Matl','Roof Sheathing','Roof Sheathing'/ Roof_Sheathing_Main
&OBST XB=22.40,24.60,6.00,6.50,4.10,4.20, SURF_IDS='Overhang Matl','Roof Sheathing','Roof Sheathing02'/ Roof_Sheathing_Main
&OBST XB=16.40,22.00,6.00,6.50,4.10,4.20, SURF_IDS='Overhang Matl','Roof Sheathing','Roof Sheathing'/ Roof_Sheathing_Main
&OBST XB=4.30,9.30,6.00,6.50,4.10,4.20, SURF_IDS='Overhang Matl','Roof Sheathing','Roof Sheathing'/ Roof_Sheathing_Main
&OBST XB=-0.90,-0.40,0.80,11.70,2.80,2.90, SURF_IDS='Overhang Matl','Roof Sheathing','Roof Sheathing'/ Roof_Sheathing_B
&OBST XB=-0.50,0.0,1.20,10.90,2.90,3.00, SURF_IDS='Overhang Matl','Roof Sheathing','Roof Sheathing'/ Roof_Sheathing_B
&OBST XB=-0.1,0.40,1.60,10.90,3.00,3.10, SURF_IDS='Overhang Matl','Roof Sheathing','Roof Sheathing'/ Roof_Sheathing_B
&OBST XB=0.3,0.80,2.00,10.50,3.10,3.20, SURF_IDS='Overhang Matl','Roof Sheathing','Roof Sheathing'/ Roof_Sheathing_B
&OBST XB=0.70,1.20,2.40,10.10,3.20,3.30, SURF_IDS='Overhang Matl','Roof Sheathing','Roof Sheathing'/ Roof_Sheathing_B
&OBST XB=1.50,2.00,3.20,9.30,3.40,3.50, SURF_IDS='Overhang Matl','Roof Sheathing','Roof Sheathing'/ Roof_Sheathing_B
&OBST XB=2.30,2.80,4.00,8.50,3.60,3.70, SURF_IDS='Overhang Matl','Roof Sheathing','Roof Sheathing'/ Roof_Sheathing_B
&OBST XB=1.90,2.40,3.60,8.90,3.50,3.60, SURF_IDS='Overhang Matl','Roof Sheathing','Roof Sheathing'/ Roof_Sheathing_B
&OBST XB=2.70,3.20,4.40,8.10,3.70,3.80, SURF_IDS='Overhang Matl','Roof Sheathing','Roof Sheathing'/ Roof_Sheathing_B
&OBST XB=3.10,3.60,4.80,7.70,3.80,3.90, SURF_IDS='Overhang Matl','Roof Sheathing','Roof Sheathing'/ Roof_Sheathing_B
&OBST XB=3.50,4.00,5.20,7.30,3.90,4.00, SURF_IDS='Overhang Matl','Roof Sheathing','Roof Sheathing'/ Roof_Sheathing_B
&OBST XB=3.90,4.40,5.60,6.90,4.00,4.10, SURF_IDS='Overhang Matl','Roof Sheathing','Roof Sheathing'/ Roof_Sheathing_B
&OBST XB=1.10,1.60,2.80,9.70,3.30,3.40, SURF_IDS='Overhang Matl','Roof Sheathing','Roof Sheathing'/ Roof_Sheathing_B
&OBST XB=-1.30,-0.80,0.40,12.10,2.70,2.80, SURF_IDS='Overhang Matl','Roof Sheathing','Roof Sheathing'/ Roof_Sheathing_B
&OBST XB=24.80,25.00,4.60,5.20,3.90,4.00, SURF_IDS='Overhang Matl','Roof Sheathing','Roof Sheathing'/ Roof_Sheathing_D
&OBST XB=24.90,30.10,0.2000,0.70,2.90,3.00, SURF_IDS='Overhang Matl','Roof Sheathing','Roof Sheathing'/ Roof_Sheathing_D
&OBST XB=24.90,29.60,0.60,1.10,3.00,3.10, SURF_IDS='Overhang Matl','Roof Sheathing','Roof Sheathing'/ Roof_Sheathing_D
&OBST XB=24.90,29.20,1.00,1.50,3.10,3.20, SURF_IDS='Overhang Matl','Roof Sheathing','Roof Sheathing'/ Roof_Sheathing_D
```

```
&OBST XB=24.90,28.80,1.40,1.90,3.20,3.30, SURF_IDS='Overhang Matl','Roof Sheathing','Roof Sheathing'/ Roof_Sheathing_D
&OBST XB=24.90,28.40,1.80,2.30,3.30,3.40, SURF_IDS='Overhang Matl','Roof Sheathing','Roof Sheathing'/ Roof_Sheathing_D
&OBST XB=24.90,28.00,2.20,2.70,3.40,3.50, SURF_IDS='Overhang Matl','Roof Sheathing','Roof Sheathing'/ Roof_Sheathing_D
&OBST XB=24.90,27.60,2.60,3.10,3.50,3.60, SURF_IDS='Overhang Matl','Roof Sheathing','Roof Sheathing'/ Roof_Sheathing_D
&OBST XB=24.90,26.80,3.40,3.90,3.70,3.80, SURF_IDS='Overhang Matl','Roof Sheathing','Roof Sheathing'/ Roof_Sheathing_D
&OBST XB=24.90,26.40,3.80,4.30,3.80,3.90, SURF_IDS='Overhang Matl','Roof Sheathing','Roof Sheathing'/ Roof_Sheathing_D
&OBST XB=24.90,26.00,4.20,4.70,3.90,4.00, SURF_IDS='Overhang Matl','Roof Sheathing','Roof Sheathing'/ Roof_Sheathing_D
&OBST XB=24.90,30.80,-0.60,-0.1,2.70,2.80, SURF_IDS='Overhang Matl','Roof Sheathing','Roof Sheathing'/ Roof_Sheathing_D
&OBST XB=24.90,27.20,3.00,3.50,3.60,3.70, SURF_IDS='Overhang Matl','Roof Sheathing','Roof Sheathing'/ Roof_Sheathing_D
&OBST XB=24.80,25.20,4.60,5.70,4.00,4.10, SURF_IDS='Overhang Matl','Roof Sheathing','Roof Sheathing'/ Roof_Sheathing_D
&OBST XB=24.80,25.00,4.20,4.80,3.80,3.90, SURF_IDS='Overhang Matl','Roof Sheathing','Roof Sheathing'/ Roof_Sheathing_D
&OBST XB=24.80,25.00,3.80,4.40,3.70,3.80, SURF_IDS='Overhang Matl','Roof Sheathing','Roof Sheathing'/ Roof_Sheathing_D
&OBST XB=24.80,25.00,3.40,4.00,3.60,3.70, SURF_IDS='Overhang Matl','Roof Sheathing','Roof Sheathing'/ Roof_Sheathing_D
&OBST XB=24.80,25.00,3.00,3.60,3.50,3.60, SURF_IDS='Overhang Matl','Roof Sheathing','Roof Sheathing'/ Roof_Sheathing_D
&OBST XB=24.80,25.00,2.60,3.20,3.40,3.50, SURF_IDS='Overhang Matl','Roof Sheathing','Roof Sheathing'/ Roof_Sheathing_D
&OBST XB=24.80,25.00,2.20,2.80,3.30,3.40, SURF_IDS='Overhang Matl','Roof Sheathing','Roof Sheathing'/ Roof_Sheathing_D
&OBST XB=24.80,25.00,1.80,2.40,3.20,3.30, SURF_IDS='Overhang Matl','Roof Sheathing','Roof Sheathing'/ Roof_Sheathing_D
&OBST XB=24.80,25.00,1.40,2.00,3.10,3.20, SURF_IDS='Overhang Matl','Roof Sheathing','Roof Sheathing'/ Roof_Sheathing_D
&OBST XB=24.80,25.00,1.00,1.60,3.00,3.10, SURF_IDS='Overhang Matl','Roof Sheathing','Roof Sheathing'/ Roof_Sheathing_D
&OBST XB=24.80,25.00,0.60,1.20,2.90,3.00, SURF_IDS='Overhang Matl','Roof Sheathing','Roof Sheathing'/ Roof_Sheathing_D
&OBST XB=24.80,25.00,0.2000,0.80,2.80,2.90, SURF_IDS='Overhang Matl','Roof Sheathing','Roof Sheathing'/ Roof_Sheathing_D
&OBST XB=24.90,25.10,-0.2000,0.40,2.70,2.80, SURF_IDS='Overhang Matl','Roof Sheathing','Roof Sheathing'/ Roof_Sheathing_D
&OBST XB=24.90,30.400,-0.2000,0.3,2.80,2.90, SURF_IDS='Overhang Matl','Roof Sheathing','Roof Sheathing'/ Roof_Sheathing_D
&OBST XB=30.30,30.80,-0.1,5.40,2.70,2.80, SURF_IDS='Overhang Matl','Roof Sheathing','Roof Sheathing'/ Roof_Sheathing_D
&OBST XB=25.90,26.40,4.30,7.00,3.80,3.90, SURF_IDS='Overhang Matl','Roof Sheathing','Roof Sheathing'/ Roof_Sheathing_D
&OBST XB=26.30,26.80,3.90,6.80,3.70,3.80, SURF_IDS='Overhang Matl','Roof Sheathing','Roof Sheathing'/ Roof_Sheathing_D
&OBST XB=27.10,27.60,3.10,6.50,3.50,3.60, SURF_IDS='Overhang Matl','Roof Sheathing','Roof Sheathing'/ Roof_Sheathing_D
&OBST XB=26.70,27.20,3.40,6.60,3.60,3.70, SURF_IDS='Overhang Matl','Roof Sheathing','Roof Sheathing'/ Roof_Sheathing_D
&OBST XB=27.90,28.40,2.30,6.20,3.30,3.40, SURF_IDS='Overhang Matl','Roof Sheathing','Roof Sheathing'/ Roof_Sheathing_D
&OBST XB=27.50,28.00,2.70,6.30,3.40,3.50, SURF_IDS='Overhang Matl','Roof Sheathing','Roof Sheathing'/ Roof_Sheathing_D
&OBST XB=28.30,28.80,1.90,6.00,3.20,3.30, SURF_IDS='Overhang Matl','Roof Sheathing','Roof Sheathing'/ Roof_Sheathing_D
&OBST XB=28.70,29.20,1.50,5.80,3.10,3.20, SURF_IDS='Overhang Matl','Roof Sheathing','Roof Sheathing'/ Roof_Sheathing_D
&OBST XB=29.10,29.60,1.10,5.70,3.00,3.10, SURF_IDS='Overhang Matl','Roof Sheathing','Roof Sheathing'/ Roof_Sheathing_D
&OBST XB=29.50,30.10,0.70,5.50,2.90,3.00, SURF_IDS='Overhang Matl','Roof Sheathing','Roof Sheathing'/ Roof_Sheathing_D
&OBST XB=29.90,30.400,0.3,5.40,2.80,2.90, SURF_IDS='Overhang Matl','Roof Sheathing','Roof Sheathing'/ Roof_Sheathing_D
&OBST XB=3.10,14.80,15.40,16.00,2.80,2.90, SURF_IDS='Overhang Matl','Roof Sheathing','Roof Sheathing'/ Roof_Sheathing_LAdd
&OBST XB=2.60,15.30,15.90,16.50,2.70,2.80, SURF_IDS='Overhang Matl','Roof Sheathing','Roof Sheathing'/ Roof_Sheathing_LAdd
&OBST XB=3.70,14.30,14.90,15.50,2.90,3.00, SURF_IDS='Overhang Matl','Roof Sheathing','Roof Sheathing'/ Roof_Sheathing_LAdd
&OBST XB=4.10,13.80,14.40,15.00,3.00,3.10, SURF_IDS='Overhang Matl','Roof Sheathing','Roof Sheathing'/ Roof_Sheathing_LAdd
&OBST XB=4.60,13.30,13.80,14.50,3.10,3.20, SURF_IDS='Overhang Matl','Roof Sheathing','Roof Sheathing'/ Roof_Sheathing_LAdd
&OBST XB=5.10,12.80,13.20,13.85,3.20,3.30, SURF_IDS='Overhang Matl','Roof Sheathing','Roof Sheathing'/ Roof_Sheathing_LAdd
&OBST XB=5.60,12.30,12.70,13.30,3.30,3.40, SURF_IDS='Overhang Matl','Roof Sheathing','Roof Sheathing'/ Roof_Sheathing_LAdd
&OBST XB=6.10,11.80,12.20,12.80,3.40,3.50, SURF_IDS='Overhang Matl','Roof Sheathing','Roof Sheathing'/ Roof_Sheathing_LAdd
&OBST XB=6.60,11.30,11.80,12.30,3.50,3.60, SURF_IDS='Overhang Matl','Roof Sheathing','Roof Sheathing'/ Roof_Sheathing_LAdd
&OBST XB=7.10,10.80,11.30,11.90,3.60,3.70, SURF_IDS='Overhang Matl','Roof Sheathing','Roof Sheathing'/ Roof_Sheathing_LAdd
&OBST XB=7.40,10.50,10.90,11.40,3.70,3.80, SURF_IDS='Overhang Matl','Roof Sheathing','Roof Sheathing'/ Roof_Sheathing_LAdd
&OBST XB=7.70,10.20,10.50,11.00,3.80,3.90, SURF_IDS='Overhang Matl','Roof Sheathing','Roof Sheathing'/ Roof_Sheathing_LAdd
&OBST XB=8.00,9.90,10.10,10.60,3.90,4.00, SURF_IDS='Overhang Matl','Roof Sheathing','Roof Sheathing'/ Roof_Sheathing_LAdd
&OBST XB=8.30,9.60,9.70,10.20,4.00,4.10, SURF_IDS='Overhang Matl','Roof Sheathing','Roof Sheathing'/ Roof_Sheathing_LAdd
&OBST XB=8.60,9.30,6.50,9.80,4.10,4.20, SURF_IDS='Overhang Matl','Roof Sheathing','Roof Sheathing'/ Roof_Sheathing_LAdd
&OBST XB=8.30,9.60,6.90,9.70,4.00,4.10, SURF_IDS='Overhang Matl','Roof Sheathing','Roof Sheathing'/ Roof_Sheathing_LAdd
&OBST XB=9.50,9.90,7.30,10.10,3.90,4.00, SURF_IDS='Overhang Matl','Roof Sheathing','Roof Sheathing'/ Roof_Sheathing_LAdd
&OBST XB=10.10,10.50,8.10,10.90,3.70,3.80, SURF_IDS='Overhang Matl','Roof Sheathing','Roof Sheathing'/ Roof_Sheathing_LAdd
&OBST XB=10.400,10.80,8.50,11.30,3.60,3.70, SURF_IDS='Overhang Matl','Roof Sheathing','Roof Sheathing'/ Roof_Sheathing_LAdd
&OBST XB=14.80,15.30,13.60,15.90,2.70,2.80, SURF_IDS='Overhang Matl','Roof Sheathing','Roof Sheathing'/ Roof_Sheathing_LAdd
&OBST XB=10.80,11.30,8.90,11.80,3.50,3.60, SURF_IDS='Overhang Matl','Roof Sheathing','Roof Sheathing'/ Roof_Sheathing_LAdd
&OBST XB=11.30,11.80,9.30,12.20,3.40,3.50, SURF_IDS='Overhang Matl','Roof Sheathing','Roof Sheathing'/ Roof_Sheathing_LAdd
&OBST XB=11.80,12.30,9.70,12.70,3.30,3.40, SURF_IDS='Overhang Matl','Roof Sheathing','Roof Sheathing'/ Roof_Sheathing_LAdd
&OBST XB=12.30,12.80,10.10,13.20,3.20,3.30, SURF_IDS='Overhang Matl','Roof Sheathing','Roof Sheathing'/ Roof_Sheathing_LAdd
&OBST XB=12.80,13.30,10.50,13.80,3.10,3.20, SURF_IDS='Overhang Matl','Roof Sheathing','Roof Sheathing'/ Roof_Sheathing_LAdd
&OBST XB=13.30,13.80,10.90,14.40,3.00,3.10, SURF_IDS='Overhang Matl','Roof Sheathing','Roof Sheathing'/ Roof_Sheathing_LAdd
&OBST XB=13.80,14.30,11.80,14.90,2.90,3.00, SURF_IDS='Overhang Matl','Roof Sheathing','Roof Sheathing'/ Roof_Sheathing_LAdd
&OBST XB=14.30,14.80,12.80,15.40,2.80,2.90, SURF_IDS='Overhang Matl','Roof Sheathing','Roof Sheathing'/ Roof_Sheathing_LAdd
&OBST XB=8.00,8.40,7.30,10.10,3.90,4.00, SURF_IDS='Overhang Matl','Roof Sheathing','Roof Sheathing'/ Roof_Sheathing_LAdd
&OBST XB=9.80,10.20,7.70,10.50,3.80,3.90, SURF_IDS='Overhang Matl','Roof Sheathing','Roof Sheathing'/ Roof_Sheathing_LAdd
&OBST XB=7.70,8.10,7.70,10.50,3.80,3.90, SURF_IDS='Overhang Matl','Roof Sheathing','Roof Sheathing'/ Roof_Sheathing_LAdd
&OBST XB=7.40,7.80,8.10,10.90,3.70,3.80, SURF_IDS='Overhang Matl','Roof Sheathing','Roof Sheathing'/ Roof_Sheathing_LAdd
&OBST XB=7.10,7.50,8.50,11.30,3.60,3.70, SURF_IDS='Overhang Matl','Roof Sheathing','Roof Sheathing'/ Roof_Sheathing_LAdd
&OBST XB=6.60,7.20,8.90,11.80,3.50,3.60, SURF_IDS='Overhang Matl','Roof Sheathing','Roof Sheathing'/ Roof_Sheathing_LAdd
&OBST XB=6.10,6.70,9.30,12.20,3.40,3.50, SURF_IDS='Overhang Matl','Roof Sheathing','Roof Sheathing'/ Roof_Sheathing_LAdd
&OBST XB=5.60,6.20,9.70,12.70,3.30,3.40, SURF_IDS='Overhang Matl','Roof Sheathing','Roof Sheathing'/ Roof_Sheathing_LAdd
&OBST XB=5.10,5.70,10.10,13.20,3.20,3.30, SURF_IDS='Overhang Matl','Roof Sheathing','Roof Sheathing'/ Roof_Sheathing_LAdd
&OBST XB=4.60,5.20,10.50,13.80,3.10,3.20, SURF_IDS='Overhang Matl','Roof Sheathing','Roof Sheathing'/ Roof_Sheathing_LAdd
&OBST XB=4.10,4.70,10.90,14.40,3.00,3.10, SURF_IDS='Overhang Matl','Roof Sheathing','Roof Sheathing'/ Roof_Sheathing_LAdd
&OBST XB=3.60,4.20,11.50,15.50,2.90,3.00, SURF_IDS='Overhang Matl','Roof Sheathing','Roof Sheathing'/ Roof_Sheathing_LAdd
&OBST XB=3.10,3.70,11.60,15.40,2.80,2.90, SURF_IDS='Overhang Matl','Roof Sheathing','Roof Sheathing'/ Roof_Sheathing_LAdd
&OBST XB=2.60,3.40,12.10,15.90,2.70,2.80, SURF_IDS='Overhang Matl','Roof Sheathing','Roof Sheathing'/ Roof_Sheathing_LAdd
&OBST XB=20.20,30.80,17.90,18.50,2.70,2.80, SURF_IDS='Overhang Matl','Roof Sheathing','Roof Sheathing'/ Roof_Sheathing_ExAdd
&OBST XB=20.50,30.400,17.40,18.00,2.80,2.90, SURF_IDS='Overhang Matl','Roof Sheathing','Roof Sheathing'/ Roof_Sheathing_ExAdd
&OBST XB=20.80,30.10,17.00,17.50,2.90,3.00, SURF_IDS='Overhang Matl','Roof Sheathing','Roof Sheathing'/ Roof_Sheathing_ExAdd
&OBST XB=21.10,29.60,16.50,17.10,3.00,3.10, SURF_IDS='Overhang Matl','Roof Sheathing','Roof Sheathing'/ Roof_Sheathing_ExAdd
&OBST XB=21.40,29.20,16.10,16.60,3.10,3.20, SURF_IDS='Overhang Matl','Roof Sheathing','Roof Sheathing'/ Roof_Sheathing_ExAdd
&OBST XB=21.70,28.80,15.60,16.20,3.20,3.30, SURF_IDS='Overhang Matl','Roof Sheathing','Roof Sheathing'/ Roof_Sheathing_ExAdd
&OBST XB=22.00,28.40,15.20,15.70,3.30,3.40, SURF_IDS='Overhang Matl','Roof Sheathing','Roof Sheathing'/ Roof_Sheathing_ExAdd
&OBST XB=22.30,28.00,14.70,15.30,3.40,3.50, SURF_IDS='Overhang Matl','Roof Sheathing','Roof Sheathing'/ Roof_Sheathing_ExAdd
&OBST XB=23.20,27.20,13.80,14.40,3.60,3.70, SURF_IDS='Overhang Matl','Roof Sheathing','Roof Sheathing'/ Roof_Sheathing_ExAdd
&OBST XB=23.50,26.80,13.40,13.90,3.70,3.80, SURF_IDS='Overhang Matl','Roof Sheathing','Roof Sheathing'/ Roof_Sheathing_ExAdd
&OBST XB=24.90,25.70,11.80,12.70,4.00,4.10, SURF_IDS='Overhang Matl','Roof Sheathing','Roof Sheathing'/ Roof_Sheathing_ExAdd
&OBST XB=23.50,23.90,8.10,13.40,3.70,3.80, SURF_IDS='Overhang Matl','Roof Sheathing','Roof Sheathing'/ Roof_Sheathing_ExAdd
&OBST XB=23.20,23.60,8.50,13.80,3.60,3.70, SURF_IDS='Overhang Matl','Roof Sheathing','Roof Sheathing'/ Roof_Sheathing_ExAdd
&OBST XB=22.70,23.30,8.90,14.30,3.50,3.60, SURF_IDS='Overhang Matl','Roof Sheathing','Roof Sheathing'/ Roof_Sheathing_ExAdd
&OBST XB=22.30,22.70,9.30,14.70,3.40,3.50, SURF_IDS='Overhang Matl','Roof Sheathing','Roof Sheathing'/ Roof_Sheathing_ExAdd
&OBST XB=22.00,22.40,9.70,15.20,3.30,3.40, SURF_IDS='Overhang Matl','Roof Sheathing','Roof Sheathing'/ Roof_Sheathing_ExAdd
&OBST XB=21.70,22.10,10.10,15.60,3.20,3.30, SURF_IDS='Overhang Matl','Roof Sheathing','Roof Sheathing'/ Roof_Sheathing_ExAdd
&OBST XB=21.40,21.80,10.50,16.10,3.10,3.20, SURF_IDS='Overhang Matl','Roof Sheathing','Roof Sheathing'/ Roof_Sheathing_ExAdd
```

```
&OBST XB=21.11,21.51,10.90,16.50,3.00,3.10, SURF_IDS='Overhang Matl','Roof Sheathing','Roof Sheathing'/ Roof_Sheathing_ExAdd
&OBST XB=20.82,21.22,11.80,17.00,2.90,3.00, SURF_IDS='Overhang Matl','Roof Sheathing','Roof Sheathing'/ Roof_Sheathing_ExAdd
&OBST XB=20.50,20.90,12.80,17.40,2.80,2.90, SURF_IDS='Overhang Matl','Roof Sheathing','Roof Sheathing'/ Roof_Sheathing_ExAdd
&OBST XB=20.20,20.70,13.60,17.90,2.70,2.80, SURF_IDS='Overhang Matl','Roof Sheathing','Roof Sheathing'/ Roof_Sheathing_ExAdd
&OBST XB=30.00,30.400,9.30,17.40,2.80,2.90, SURF_IDS='Overhang Matl','Roof Sheathing','Roof Sheathing'/ Roof_Sheathing_ExAdd
&OBST XB=29.50,30.10,9.20,17.00,2.90,3.00, SURF_IDS='Overhang Matl','Roof Sheathing','Roof Sheathing'/ Roof_Sheathing_ExAdd
&OBST XB=29.10,29.60,9.00,16.50,3.00,3.10, SURF_IDS='Overhang Matl','Roof Sheathing','Roof Sheathing'/ Roof_Sheathing_ExAdd
&OBST XB=28.70,29.20,8.90,16.10,3.10,3.20, SURF_IDS='Overhang Matl','Roof Sheathing','Roof Sheathing'/ Roof_Sheathing_ExAdd
&OBST XB=23.80,24.30,7.70,13.10,3.80,3.90, SURF_IDS='Overhang Matl','Roof Sheathing','Roof Sheathing'/ Roof_Sheathing_ExAdd
&OBST XB=24.20,24.70,7.30,12.20,3.90,4.00, SURF_IDS='Overhang Matl','Roof Sheathing','Roof Sheathing'/ Roof_Sheathing_ExAdd
&OBST XB=28.30,28.80,8.70,15.60,3.20,3.30, SURF_IDS='Overhang Matl','Roof Sheathing','Roof Sheathing'/ Roof_Sheathing_ExAdd
&OBST XB=27.90,28.40,8.50,15.20,3.30,3.40, SURF_IDS='Overhang Matl','Roof Sheathing','Roof Sheathing'/ Roof_Sheathing_ExAdd
&OBST XB=27.50,28.00,8.30,14.70,3.40,3.50, SURF_IDS='Overhang Matl','Roof Sheathing','Roof Sheathing'/ Roof_Sheathing_ExAdd
&OBST XB=27.10,27.60,8.20,14.30,3.50,3.60, SURF_IDS='Overhang Matl','Roof Sheathing','Roof Sheathing'/ Roof_Sheathing_ExAdd
&OBST XB=26.70,27.20,8.10,13.80,3.60,3.70, SURF_IDS='Overhang Matl','Roof Sheathing','Roof Sheathing'/ Roof_Sheathing_ExAdd
&OBST XB=26.30,26.80,7.90,13.40,3.70,3.80, SURF_IDS='Overhang Matl','Roof Sheathing','Roof Sheathing'/ Roof_Sheathing_ExAdd
&OBST XB=25.90,26.40,7.70,13.10,3.80,3.90, SURF_IDS='Overhang Matl','Roof Sheathing','Roof Sheathing'/ Roof_Sheathing_ExAdd
&OBST XB=22.70,27.60,14.30,14.80,3.50,3.60, SURF_IDS='Overhang Matl','Roof Sheathing','Roof Sheathing'/ Roof_Sheathing_ExAdd
&OBST XB=24.60,24.90,6.90,12.70,4.00,4.10, SURF_IDS='Overhang Matl','Roof Sheathing','Roof Sheathing'/ Roof_Sheathing_ExAdd
&OBST XB=23.80,26.40,12.70,13.50,3.80,3.90, SURF_IDS='Overhang Matl','Roof Sheathing','Roof Sheathing'/ Roof_Sheathing_ExAdd
&OBST XB=24.20,26.00,12.10,13.10,3.90,4.00, SURF_IDS='Overhang Matl','Roof Sheathing','Roof Sheathing'/ Roof_Sheathing_ExAdd
&OBST XB=25.20,25.70,4.60,11.80,4.00,4.10, SURF_IDS='Overhang Matl','Roof Sheathing','Roof Sheathing'/ Roof_Sheathing_ExAdd
&OBST XB=24.80,25.30,5.00,12.30,4.10,4.20, SURF_IDS='Overhang Matl','Roof Sheathing','Roof Sheathing'/ Roof_Sheathing_ExAdd
&OBST XB=25.50,26.00,4.70,7.10,3.90,4.00, SURF_IDS='Overhang Matl','Roof Sheathing','Roof Sheathing'/ Roof_Sheathing_ExAdd
&OBST XB=25.50,26.00,7.60,12.70,3.90,4.00, SURF_IDS='Overhang Matl','Roof Sheathing','Roof Sheathing'/ Roof_Sheathing_ExAdd[1]
&OBST XB=30.400,30.80,9.30,17.90,2.70,2.80, SURF_IDS='Overhang Matl','Roof Sheathing','Roof Sheathing'/ Roof_Sheathing_ExAdd[1]
&OBST XB=33.30,33.60,0.40,5.30,2.80,2.90, SURF_IDS='Overhang Matl','Roof Sheathing','Roof Sheathing'/ BathAdditionRoof
&OBST XB=33.20,33.50,0.40,5.50,2.90,3.00, SURF_IDS='Overhang Matl','Roof Sheathing','Roof Sheathing'/ BathAdditionRoof
&OBST XB=33.10,33.40,0.40,5.50,3.00,3.10, SURF_IDS='Overhang Matl','Roof Sheathing','Roof Sheathing'/ BathAdditionRoof
&OBST XB=33.00,33.30,0.40,5.70,3.10,3.20, SURF_IDS='Overhang Matl','Roof Sheathing','Roof Sheathing'/ BathAdditionRoof
&OBST XB=32.90,33.20,0.40,5.80,3.20,3.30, SURF_IDS='Overhang Matl','Roof Sheathing','Roof Sheathing'/ BathAdditionRoof
&OBST XB=32.80,33.10,0.40,6.00,3.30,3.40, SURF_IDS='Overhang Matl','Roof Sheathing','Roof Sheathing'/ BathAdditionRoof
&OBST XB=32.70,33.00,0.40,6.20,3.40,3.50, SURF_IDS='Overhang Matl','Roof Sheathing','Roof Sheathing'/ BathAdditionRoof
&OBST XB=32.60,32.90,0.40,6.30,3.50,3.60, SURF_IDS='Overhang Matl','Roof Sheathing','Roof Sheathing'/ BathAdditionRoof
&OBST XB=32.50,32.80,0.40,6.50,3.60,3.70, SURF_IDS='Overhang Matl','Roof Sheathing','Roof Sheathing'/ BathAdditionRoof
&OBST XB=32.40,32.70,0.40,6.60,3.70,3.80, SURF_IDS='Overhang Matl','Roof Sheathing','Roof Sheathing'/ BathAdditionRoof
&OBST XB=32.30,32.60,0.40,6.80,3.80,3.90, SURF_IDS='Overhang Matl','Roof Sheathing','Roof Sheathing'/ BathAdditionRoof
&OBST XB=33.40,33.70,0.40,5.10,2.70,2.80, SURF_IDS='Overhang Matl','Roof Sheathing','Roof Sheathing'/ BathAdditionRoof
&OBST XB=32.20,32.30,0.50,7.10,3.90,4.10, SURF_IDS='Overhang Matl','Roof Sheathing','Roof Sheathing'/ BathAdditionRoof
&OBST XB=32.20,32.50,0.40,7.00,3.90,4.00, SURF_IDS='Overhang Matl','Roof Sheathing','Roof Sheathing'/ BathAdditionRoof
&OBST XB=30.80,31.10,0.40,5.30,2.80,2.90, SURF_IDS='Overhang Matl','Roof Sheathing','Roof Sheathing'/ BathAdditionRoof[1]
&OBST XB=30.90,31.20,0.40,5.50,2.90,3.00, SURF_IDS='Overhang Matl','Roof Sheathing','Roof Sheathing'/ BathAdditionRoof[1]
&OBST XB=31.00,31.30,0.40,5.50,3.00,3.10, SURF_IDS='Overhang Matl','Roof Sheathing','Roof Sheathing'/ BathAdditionRoof[1]
&OBST XB=31.10,31.40,0.40,5.70,3.10,3.20, SURF_IDS='Overhang Matl','Roof Sheathing','Roof Sheathing'/ BathAdditionRoof[1]
&OBST XB=31.20,31.50,0.40,5.80,3.20,3.30, SURF_IDS='Overhang Matl','Roof Sheathing','Roof Sheathing'/ BathAdditionRoof[1]
&OBST XB=31.30,31.60,0.40,6.00,3.30,3.40, SURF_IDS='Overhang Matl','Roof Sheathing','Roof Sheathing'/ BathAdditionRoof[1]
&OBST XB=31.40,31.70,0.40,6.20,3.40,3.50, SURF_IDS='Overhang Matl','Roof Sheathing','Roof Sheathing'/ BathAdditionRoof[1]
&OBST XB=31.50,31.80,0.40,6.30,3.50,3.60, SURF_IDS='Overhang Matl','Roof Sheathing','Roof Sheathing'/ BathAdditionRoof[1]
&OBST XB=31.60,31.90,0.40,6.50,3.60,3.70, SURF_IDS='Overhang Matl','Roof Sheathing','Roof Sheathing'/ BathAdditionRoof[1]
&OBST XB=31.70,32.00,0.40,6.60,3.70,3.80, SURF_IDS='Overhang Matl','Roof Sheathing','Roof Sheathing'/ BathAdditionRoof[1]
&OBST XB=31.80,32.10,0.40,6.80,3.80,3.90, SURF_IDS='Overhang Matl','Roof Sheathing','Roof Sheathing'/ BathAdditionRoof[1]
&OBST XB=30.70,31.00,0.40,5.10,2.70,2.80, SURF_IDS='Overhang Matl','Roof Sheathing','Roof Sheathing'/ BathAdditionRoof[1]
&OBST XB=31.90,32.20,0.40,7.00,3.90,4.00, SURF_IDS='Overhang Matl','Roof Sheathing','Roof Sheathing'/ BathAdditionRoof[1]
&OBST XB=25.70,33.70,7.10,7.60,4.00,4.10, SURF_IDS='Overhang Matl','Roof Sheathing','Roof Sheathing'/ Roof_Sheathing_Steam
&OBST XB=30.80,33.70,9.70,9.90,2.40,2.60, SURF_IDS='Overhang Matl','Roof Sheathing','Roof Sheathing'/ Roof_Sheathing_Steam
&OBST XB=30.80,33.70,9.50,9.80,2.60,2.70, SURF_IDS='Overhang Matl','Roof Sheathing','Roof Sheathing'/ Roof_Sheathing_Steam
&OBST XB=30.80,33.70,9.40,9.60,2.70,2.80, SURF_IDS='Overhang Matl','Roof Sheathing','Roof Sheathing'/ Roof_Sheathing_Steam
&OBST XB=30.400,33.70,9.30,9.40,2.80,2.90, SURF_IDS='Overhang Matl','Roof Sheathing','Roof Sheathing'/ Roof_Sheathing_Steam
&OBST XB=30.10,33.70,9.10,9.30,2.90,3.00, SURF_IDS='Overhang Matl','Roof Sheathing','Roof Sheathing'/ Roof_Sheathing_Steam
&OBST XB=29.60,33.70,8.90,9.20,3.00,3.10, SURF_IDS='Overhang Matl','Roof Sheathing','Roof Sheathing'/ Roof_Sheathing_Steam
&OBST XB=29.20,33.70,8.81,9.00,3.10,3.20, SURF_IDS='Overhang Matl','Roof Sheathing','Roof Sheathing'/ Roof_Sheathing_Steam
&OBST XB=28.80,33.70,8.60,8.90,3.20,3.30, SURF_IDS='Overhang Matl','Roof Sheathing','Roof Sheathing'/ Roof_Sheathing_Steam
&OBST XB=28.40,33.70,8.40,8.70,3.30,3.40, SURF_IDS='Overhang Matl','Roof Sheathing','Roof Sheathing'/ Roof_Sheathing_Steam
&OBST XB=28.00,33.70,8.30,8.50,3.40,3.50, SURF_IDS='Overhang Matl','Roof Sheathing','Roof Sheathing'/ Roof_Sheathing_Steam
&OBST XB=27.60,33.70,8.10,8.30,3.50,3.60, SURF_IDS='Overhang Matl','Roof Sheathing','Roof Sheathing'/ Roof_Sheathing_Steam
&OBST XB=27.20,33.70,8.00,8.20,3.60,3.70, SURF_IDS='Overhang Matl','Roof Sheathing','Roof Sheathing'/ Roof_Sheathing_Steam
&OBST XB=26.80,33.70,7.80,8.10,3.70,3.80, SURF_IDS='Overhang Matl','Roof Sheathing','Roof Sheathing'/ Roof_Sheathing_Steam
&OBST XB=26.40,33.70,7.70,7.90,3.80,3.90, SURF_IDS='Overhang Matl','Roof Sheathing','Roof Sheathing'/ Roof_Sheathing_Steam
&OBST XB=26.00,33.70,7.50,7.70,3.90,4.00, SURF_IDS='Overhang Matl','Roof Sheathing','Roof Sheathing'/ Roof_Sheathing_Steam
&OBST XB=30.80,33.70,4.80,5.30,2.70,2.80, SURF_IDS='Overhang Matl','Roof Sheathing','Roof Sheathing'/ Roof_Sheathing_Steam
&OBST XB=30.400,33.70,5.30,5.40,2.80,2.90, SURF_IDS='Overhang Matl','Roof Sheathing','Roof Sheathing'/ Roof_Sheathing_Steam
&OBST XB=30.10,33.70,5.40,5.60,2.90,3.00, SURF_IDS='Overhang Matl','Roof Sheathing','Roof Sheathing'/ Roof_Sheathing_Steam
&OBST XB=29.60,33.70,5.50,5.80,3.00,3.10, SURF_IDS='Overhang Matl','Roof Sheathing','Roof Sheathing'/ Roof_Sheathing_Steam
&OBST XB=29.20,33.70,5.70,5.89,3.10,3.20, SURF_IDS='Overhang Matl','Roof Sheathing','Roof Sheathing'/ Roof_Sheathing_Steam
&OBST XB=28.80,33.70,5.80,6.10,3.20,3.30, SURF_IDS='Overhang Matl','Roof Sheathing','Roof Sheathing'/ Roof_Sheathing_Steam
&OBST XB=28.40,33.70,6.00,6.30,3.30,3.40, SURF_IDS='Overhang Matl','Roof Sheathing','Roof Sheathing'/ Roof_Sheathing_Steam
&OBST XB=28.00,33.70,6.20,6.40,3.40,3.50, SURF_IDS='Overhang Matl','Roof Sheathing','Roof Sheathing'/ Roof_Sheathing_Steam
&OBST XB=27.60,33.70,6.30,6.60,3.50,3.60, SURF_IDS='Overhang Matl','Roof Sheathing','Roof Sheathing'/ Roof_Sheathing_Steam
&OBST XB=27.20,33.70,6.50,6.70,3.60,3.70, SURF_IDS='Overhang Matl','Roof Sheathing','Roof Sheathing'/ Roof_Sheathing_Steam
&OBST XB=26.80,33.70,6.60,6.90,3.70,3.80, SURF_IDS='Overhang Matl','Roof Sheathing','Roof Sheathing'/ Roof_Sheathing_Steam
&OBST XB=26.40,33.70,6.80,7.00,3.80,3.90, SURF_IDS='Overhang Matl','Roof Sheathing','Roof Sheathing'/ Roof_Sheathing_Steam
&OBST XB=26.00,33.70,7.00,7.20,3.90,4.00, SURF_IDS='Overhang Matl','Roof Sheathing','Roof Sheathing'/ Roof_Sheathing_Steam
&OBST XB=20.20,20.30,13.60,17.40,2.40,2.50, RGB=153,102,0, SURF_ID='Overhang Matl'/ Roof_Fascia
&OBST XB=-1.30,2.60,12.00,12.10,2.50,2.70, RGB=153,102,0, SURF_ID='Wood Joists02'/ Roof_Fascia
&OBST XB=24.90,30.80,-0.60,-0.50,2.50,2.70, RGB=153,102,0, SURF_ID='Wood Joists02'/ Roof_FasciaA1
&OBST XB=24.90,25.00,-0.50,0.40,2.50,2.70, RGB=153,102,0, SURF_ID='Wood Joists02'/ Roof_FasciaA2
&OBST XB=-1.30,24.90,0.40,0.50,2.50,2.70, RGB=153,102,0, SURF_ID='Wood Joists02'/ Roof_FasciaA3
&OBST XB=-1.30,-1.20,0.40,12.00,2.50,2.70, RGB=153,102,0, SURF_ID='Wood Joists02'/ Roof_FasciaB
&OBST XB=30.70,30.80,-0.50,5.40,2.50,2.70, RGB=153,102,0, SURF_ID='Wood Joists02'/ Roof_FasciaD1
&OBST XB=15.30,20.30,13.50,13.60,2.50,2.70, RGB=153,102,0, SURF_ID='Wood Joists02'/ Roof_FasciaC1
&OBST XB=2.60,2.70,12.10,16.40,2.50,2.70, RGB=153,102,0, SURF_ID='Wood Joists02'/ Roof_FasciaC2
&OBST XB=2.60,15.30,16.40,16.50,2.50,2.70, RGB=153,102,0, SURF_ID='Wood Joists02'/ Roof_FasciaC3
&OBST XB=15.20,15.30,12.00,16.40,2.50,2.70, RGB=153,102,0, SURF_ID='Wood Joists02'/ Roof_FasciaC4
&OBST XB=20.20,20.30,13.60,18.40,2.50,2.70, RGB=153,102,0, SURF_ID='Wood Joists02'/ Roof_FasciaC5
&OBST XB=20.20,30.80,18.40,18.50,2.50,2.70, RGB=153,102,0, SURF_ID='Wood Joists02'/ Roof_FasciaC6
```

```
&OBST XB=30.80,33.70,0.40,0.50,2.50,2.70, RGB=153,102,0, SURF_ID='Wood Joists02'/ Roof_FasciaAD1
&OBST XB=31.00,33.40,0.40,0.50,2.70,2.80, RGB=153,102,0, SURF_ID='Roof Sheathing'/ Roof_FasciaAD2
&OBST XB=31.10,33.30,0.40,0.50,2.80,2.90, RGB=153,102,0, SURF_ID='Roof Sheathing'/ Roof_FasciaAD3
&OBST XB=31.20,33.20,0.40,0.50,2.90,3.00, RGB=153,102,0, SURF_ID='Roof Sheathing'/ Roof_FasciaAD4
&OBST XB=31.30,33.10,0.40,0.50,3.00,3.10, RGB=153,102,0, SURF_ID='Roof Sheathing'/ Roof_FasciaAD5
&OBST XB=31.40,33.00,0.40,0.50,3.10,3.20, RGB=153,102,0, SURF_ID='Roof Sheathing'/ Roof_FasciaAD6
&OBST XB=31.50,32.90,0.40,0.50,3.20,3.30, RGB=153,102,0, SURF_ID='Roof Sheathing'/ Roof_FasciaAD7
&OBST XB=31.60,32.80,0.40,0.50,3.30,3.40, RGB=153,102,0, SURF_ID='Roof Sheathing'/ Roof_FasciaAD8
&OBST XB=31.70,32.70,0.40,0.50,3.40,3.50, RGB=153,102,0, SURF_ID='Roof Sheathing'/ Roof_FasciaAD9
&OBST XB=31.80,32.60,0.40,0.50,3.50,3.90, RGB=153,102,0, SURF_ID='Roof Sheathing'/ Roof_FasciaAD10
&OBST XB=30.70,30.80,9.80,18.50,2.50,2.70, RGB=153,102,0, SURF_ID='Wood Joists02'/ Roof_FasciaD1[1]
&OBST XB=29.60,29.60,8.30,8.40,3.50,3.90, PERMIT_HOLE=.FALSE., SURF_ID='Roof Sheathing'/ RoofVent6
&OBST XB=30.00,30.00,8.30,8.40,3.50,3.90, PERMIT_HOLE=.FALSE., SURF_ID='Roof Sheathing'/ RoofVent6
&OBST XB=29.60,29.60,8.20,8.30,3.60,3.90, PERMIT_HOLE=.FALSE., SURF_ID='Roof Sheathing'/ RoofVent6
&OBST XB=30.00,30.00,8.20,8.30,3.60,3.90, PERMIT_HOLE=.FALSE., SURF_ID='Roof Sheathing'/ RoofVent6
&OBST XB=29.60,29.60,8.10,8.20,3.70,3.90, PERMIT_HOLE=.FALSE., SURF_ID='Roof Sheathing'/ RoofVent6
&OBST XB=30.00,30.00,8.10,8.20,3.70,3.90, PERMIT_HOLE=.FALSE., SURF_ID='Roof Sheathing'/ RoofVent6
&OBST XB=29.60,29.60,7.90,8.10,3.80,3.90, PERMIT_HOLE=.FALSE., SURF_ID='Roof Sheathing'/ RoofVent6
&OBST XB=30.00,30.00,7.90,8.10,3.80,3.90, PERMIT_HOLE=.FALSE., SURF_ID='Roof Sheathing'/ RoofVent6
&OBST XB=9.30,9.70,6.50,6.50,4.10,4.20, SURF_ID='Roof Sheathing'/ Center Roof VENT1
&OBST XB=9.30,9.70,6.00,6.00,4.10,4.20, SURF_ID='Roof Sheathing'/ Center Roof VENT1
&OBST XB=9.30,9.70,6.00,6.50,4.20,4.20, OUTLINE=.TRUE., SURF_ID='Vent Flow'/ TurbineBottom
&OBST XB=16.00,16.40,6.50,6.50,4.10,4.20, SURF_ID='Roof Sheathing'/ Center Roof VENT2
&OBST XB=16.00,16.40,6.00,6.00,4.10,4.20, SURF_ID='Roof Sheathing'/ Center Roof VENT2
&OBST XB=16.00,16.40,6.00,6.50,4.20,4.20, OUTLINE=.TRUE., SURF_ID='Vent Flow'/ TurbineBottom2
&OBST XB=22.00,22.40,6.50,6.50,4.10,4.20, SURF_ID='Roof Sheathing'/ Center Roof VENT3
&OBST XB=22.00,22.40,6.00,6.00,4.10,4.20, SURF_ID='Roof Sheathing'/ Center Roof VENT3
&OBST XB=22.00,22.40,6.00,6.50,4.20,4.20, OUTLINE=.TRUE., SURF_ID='Vent Flow'/ TurbineBottom3
&OBST XB=29.60,30.00,8.40,8.40,3.50,3.90, SURF_ID='Roof Sheathing'/ Center Roof VENT4
&OBST XB=29.60,30.00,7.90,7.90,3.80,3.90, SURF_ID='Roof Sheathing'/ Center Roof VENT4
&OBST XB=29.60,30.00,7.90,8.40,3.90,3.90, OUTLINE=.TRUE., SURF_ID='Vent Flow'/ TurbineBottom4
&OBST XB=9.80,10.20,11.40,11.40,3.70,3.90, PERMIT_HOLE=.FALSE., SURF_ID='Roof Sheathing'/ RoofVent5
&OBST XB=9.80,10.20,10.90,10.90,3.80,3.90, PERMIT_HOLE=.FALSE., SURF_ID='Roof Sheathing'/ RoofVent5
&OBST XB=9.80,10.20,10.90,11.40,3.90,3.90, OUTLINE=.TRUE., SURF_ID='Vent Flow'/ RoofVent5
&OBST XB=10.20,10.20,10.90,11.40,3.80,3.90, PERMIT_HOLE=.FALSE., SURF_ID='Roof Sheathing'/ RoofVent5[1]
&OBST XB=9.80,9.80,10.90,11.40,3.80,3.90, PERMIT_HOLE=.FALSE., SURF_ID='Roof Sheathing'/ RoofVent5[1]
&OBST XB=24.50,24.90,6.10,7.70,1.70,1.90, SURF_ID='Cotton'/ Closet Combustibles
&OBST XB=22.70,24.90,5.70,6.10,1.70,1.90, SURF_ID='Cotton'/ Closet Combustibles
&OBST XB=22.70,23.20,6.10,7.70,1.70,1.90, SURF_ID='Cotton'/ Closet Combustibles
&OBST XB=24.50,24.90,6.10,7.70,1.30,1.50, SURF_ID='Cotton'/ Closet Combustibles
&OBST XB=22.70,24.90,5.70,6.10,1.30,1.50, SURF_ID='Cotton'/ Closet Combustibles
&OBST XB=22.70,23.20,6.10,7.70,1.30,1.50, SURF_ID='Cotton'/ Closet Combustibles
&OBST XB=22.70,23.20,5.70,7.70,1.60,1.70, SURF_ID='shelf'/ Closet Combustibles
&OBST XB=23.20,24.90,5.70,6.10,1.60,1.70, SURF_ID='shelf'/ Closet Combustibles
&OBST XB=24.50,24.90,6.10,7.70,1.60,1.70, SURF_ID='shelf'/ Closet Combustibles
&OBST XB=22.70,23.20,5.70,7.70,1.20,1.30, SURF_ID='shelf'/ Closet Combustibles
&OBST XB=23.20,24.90,5.70,6.10,1.20,1.30, SURF_ID='shelf'/ Closet Combustibles
&OBST XB=24.50,24.90,6.10,7.70,1.20,1.30, SURF_ID='shelf'/ Closet Combustibles
&OBST XB=24.50,24.90,6.10,7.70,0.90,1.10, SURF_ID='Cotton'/ Closet Combustibles
&OBST XB=22.70,24.90,5.70,6.10,0.90,1.10, SURF_ID='Cotton'/ Closet Combustibles
&OBST XB=22.70,23.20,6.10,7.70,0.90,1.10, SURF_ID='Cotton'/ Closet Combustibles
&OBST XB=22.70,23.20,5.70,7.70,0.80,0.90, SURF_ID='shelf'/ Closet Combustibles
&OBST XB=23.20,24.90,5.70,6.10,0.80,0.90, SURF_ID='shelf'/ Closet Combustibles
&OBST XB=24.50,24.90,6.10,7.70,0.80,0.90, SURF_ID='shelf'/ Closet Combustibles
&OBST XB=24.50,24.90,6.10,7.70,0.50,0.70, SURF_ID='Cotton'/ Closet Combustibles
&OBST XB=22.70,24.90,5.70,6.10,0.50,0.70, SURF_ID='Cotton'/ Closet Combustibles
&OBST XB=22.70,23.20,6.10,7.70,0.50,0.70, SURF_ID='Cotton'/ Closet Combustibles
&OBST XB=22.70,23.20,5.70,7.70,0.40,0.50, SURF_ID='shelf'/ Closet Combustibles
&OBST XB=23.20,24.90,5.70,6.10,0.40,0.50, SURF_ID='shelf'/ Closet Combustibles
&OBST XB=24.50,24.90,6.10,7.70,0.40,0.50, SURF_ID='shelf'/ Closet Combustibles
&OBST XB=30.70,30.80,0.40,9.90,17.40,0,2.40, SURF_ID6='Gypsum','Brick','Brick','Brick','Brick','Brick'/ Wall_Exterior
&OBST XB=30.80,33.40,9.80,9.90,0,2.40, SURF_ID6='Brick','Brick','Gypsum','Brick','Brick','Brick'/ Wall_Exterior
&OBST XB=2.70,15.30,14.90,15.00,0,2.40, SURF_ID6='Brick','Brick','Gypsum','Brick','Brick','Brick'/ Wall_Exterior
&OBST XB=20.20,30.80,17.40,17.50,0,2.40, SURF_ID6='Brick','Gypsum','Brick','Brick','Brick','Brick'/ Wall_Exterior
&OBST XB=20.20,20.30,15.00,17.40,0,2.40, SURF_ID6='Brick','Gypsum','Brick','Brick','Brick','Brick'/ Wall_Exterior
&OBST XB=0.2000,0.3,2.20,12.00,0,2.40, SURF_ID6='Brick','Gypsum','Brick','Brick','Brick','Brick'/ Wall_Exterior
&OBST XB=0.2000,2.70,12.10,12.20,0,2.40, SURF_ID6='Brick','Gypsum','Brick','Brick','Brick','Brick'/ Wall_Exterior
&OBST XB=33.30,33.40,0.40,9.80,0,2.40, SURF_ID6='Gypsum','Brick','Brick','Brick','Brick','Brick'/ Wall_Exterior
&OBST XB=24.90,33.40,0.40,0.50,0,2.40, SURF_ID6='Brick','Brick','Brick','Gypsum','Brick','Brick'/ Wall_Exterior
&OBST XB=24.90,25.00,0.50,3.40,0,2.40, SURF_ID6='Brick','Gypsum','Brick','Brick','Brick','Brick'/ Wall_Exterior
&OBST XB=21.40,25.00,3.40,3.50,0,2.40, SURF_ID6='Brick','Brick','Brick','Gypsum','Brick','Brick'/ Wall_Exterior
&OBST XB=0.2000,21.40,2.10,2.20,0,2.40, SURF_ID6='Brick','Brick','Brick','Gypsum','Brick','Brick'/ Wall_Exterior
&OBST XB=21.40,21.50,2.10,3.40,0,2.40, SURF_ID6='Gypsum','Brick','Brick','Brick','Brick','Brick'/ Wall_Exterior
&OBST XB=21.40,21.50,3.50,7.70,0,2.40, SURF_ID='Gypsum'/ Wall_Exterior
&OBST XB=33.60,33.70,0.40,9.70,2.50,2.70, SURF_ID='Roof Sheathing'/ Wall_Interior
&OBST XB=30.70,30.80,5.40,7.50,2.50,2.90, PERMIT_HOLE=.FALSE., SURF_ID='Ceiling Gypsum'/ Wall_Interior
&OBST XB=30.70,30.80,5.50,7.50,2.90,3.00, PERMIT_HOLE=.FALSE., SURF_ID='Ceiling Gypsum'/ Wall_Interior
&OBST XB=30.70,30.80,5.70,7.50,3.00,3.10, PERMIT_HOLE=.FALSE., SURF_ID='Ceiling Gypsum'/ Wall_Interior
&OBST XB=30.70,30.80,5.80,7.50,3.10,3.20, PERMIT_HOLE=.FALSE., SURF_ID='Ceiling Gypsum'/ Wall_Interior
&OBST XB=30.70,30.80,6.00,7.50,3.20,3.30, PERMIT_HOLE=.FALSE., SURF_ID='Ceiling Gypsum'/ Wall_Interior
&OBST XB=30.70,30.80,6.20,7.50,3.30,3.40, PERMIT_HOLE=.FALSE., SURF_ID='Ceiling Gypsum'/ Wall_Interior
&OBST XB=30.70,30.80,6.30,7.50,3.40,3.50, PERMIT_HOLE=.FALSE., SURF_ID='Ceiling Gypsum'/ Wall_Interior
&OBST XB=30.70,30.80,6.50,7.50,3.50,3.60, PERMIT_HOLE=.FALSE., SURF_ID='Ceiling Gypsum'/ Wall_Interior
&OBST XB=30.70,30.80,6.60,7.50,3.60,3.70, PERMIT_HOLE=.FALSE., SURF_ID='Ceiling Gypsum'/ Wall_Interior
&OBST XB=30.70,30.80,6.80,7.50,3.70,3.80, PERMIT_HOLE=.FALSE., SURF_ID='Ceiling Gypsum'/ Wall_Interior
&OBST XB=30.70,30.80,7.00,7.50,3.80,3.90, PERMIT_HOLE=.FALSE., SURF_ID='Ceiling Gypsum'/ Wall_Interior
&OBST XB=30.70,30.80,7.10,7.50,3.90,4.00, PERMIT_HOLE=.FALSE., SURF_ID='Ceiling Gypsum'/ Wall_Interior
&OBST XB=33.60,33.70,7.30,7.40,2.70,4.00, PERMIT_HOLE=.FALSE., SURF_ID='Overhang Matl'/ Wall_Interior
&OBST XB=33.40,33.70,0.40,9.90,2.40,2.50, PERMIT_HOLE=.FALSE., SURF_ID='Overhang Matl'/ Wall_Interior
&OBST XB=33.30,33.60,7.50,7.60,2.50,4.00, PERMIT_HOLE=.FALSE., SURF_ID='Ceiling Gypsum'/ Wall_Interior
&OBST XB=25.00,27.90,7.70,7.80,0,2.40, SURF_ID='Gypsum'/ Wall_Interior
&OBST XB=25.00,27.90,5.60,5.70,0,2.40, SURF_ID='Gypsum'/ Wall_Interior
&OBST XB=33.30,33.70,7.60,7.60,0,2.40, SURF_ID='Gypsum'/ Wall_Interior
&OBST XB=30.70,30.80,4.70,7.50,0,2.40, SURF_ID='Gypsum'/ Wall_Interior
&OBST XB=30.70,30.80,7.60,9.90,0,2.40, SURF_ID='Gypsum'/ Wall_Interior
```

```
&OBST XB=20.20,30.70,13.60,13.70,0,2.40, SURF_ID='Gypsum'/ Wall_Interior
&OBST XB=21.80,21.90,8.90,13.60,0,2.40, SURF_ID='Gypsum'/ Wall_Interior
&OBST XB=24.90,25.00,8.90,13.70,0,2.40, SURF_ID='Gypsum'/ Wall_Interior
&OBST XB=22.80,25.00,8.90,9.00,0,2.40, SURF_ID='Gypsum'/ Wall_Interior
&OBST XB=21.20,22.80,8.90,9.00,0,2.40, SURF_ID='Gypsum'/ Wall_Interior
&OBST XB=27.80,30.70,6.30,6.40,0,2.40, SURF_ID='Gypsum'/ Wall_Interior
&OBST XB=27.80,27.90,4.80,7.70,0,2.40, SURF_ID='Gypsum'/ Wall_Interior
&OBST XB=22.60,25.00,7.70,7.80,0,2.40, SURF_ID6='Gypsum','Gypsum','Closet Walls','Gypsum','Gypsum','Gypsum'/ Wall_Interior
&OBST XB=24.90,25.00,5.60,7.80,0,2.40, SURF_ID6='Closet Walls','Gypsum','Gypsum','Gypsum','Gypsum','Gypsum'/ Wall_Interior
&OBST XB=22.70,25.00,5.60,5.70,0,2.40, SURF_ID6='Gypsum','Gypsum','Gypsum','Closet Walls','Gypsum','Gypsum'/ Wall_Interior
&OBST XB=24.10,33.30,4.70,4.80,0,2.40, SURF_ID='Gypsum'/ Wall_Interior
&OBST XB=25.10,25.20,4.70,5.60,0,2.40, SURF_ID='Gypsum'/ Wall_Interior
&OBST XB=24.00,24.10,3.50,4.80,0,2.40, SURF_ID='Gypsum'/ Wall_Interior
&OBST XB=14.60,14.70,11.40,14.90,0,2.40, SURF_ID='Gypsum'/ Wall_Interior
&OBST XB=12.60,19.50,7.60,7.70,0,2.40, SURF_ID='Gypsum'/ Wall_Interior
&OBST XB=12.60,12.70,7.70,8.90,0,2.40, SURF_ID='Gypsum'/ Wall_Interior
&OBST XB=12.70,14.70,8.80,8.90,0,2.40, SURF_ID='Gypsum'/ Wall_Interior
&OBST XB=14.60,14.70,2.20,8.80,0,2.40, SURF_ID='Gypsum'/ Wall_Interior
&OBST XB=7.10,7.20,12.70,14.90,0,2.40, SURF_ID='Gypsum'/ Wall_Interior
&OBST XB=4.70,7.10,12.70,12.80,0,2.40, SURF_ID='Gypsum'/ Wall_Interior
&OBST XB=4.70,4.80,12.80,14.90,0,2.40, SURF_ID='Gypsum'/ Wall_Interior
&OBST XB=30.70,30.80,0.50,4.80,0,2.40, SURF_ID='Gypsum'/ Wall_Interior
&OBST XB=2.70,2.80,2.20,12.00,0,2.40, SURF_ID='Gypsum'/ Wall_Interior
&OBST XB=0.3,2.70,6.50,6.60,0,2.40, SURF_ID='Gypsum'/ Wall_Interior
&OBST XB=2.80,9.20,9.90,10.00,0,2.40, SURF_ID='Gypsum'/ Wall_Interior
&OBST XB=9.10,9.20,7.90,9.90,0,2.40, SURF_ID='Gypsum'/ Wall_Interior
&OBST XB=9.20,10.60,7.90,8.00,0,2.40, SURF_ID='Gypsum'/ Wall_Interior
&OBST XB=10.10,10.20,2.20,7.90,0,2.40, SURF_ID='Gypsum'/ Wall_Interior
&OBST XB=22.60,22.70,5.60,7.70,0,2.40, SURF_ID6='Gypsum','Closet Walls','Gypsum','Gypsum','Gypsum','Gypsum'/ Wall_Interior
&OBST XB=2.70,2.80,12.00,15.00,0,2.40, SURF_ID6='Brick','Gypsum','Brick','Brick','Brick','Brick'/ Wall_Interior
&OBST XB=14.70,15.20,11.90,12.50,2.40,4.00, COLOR='GRAY 80', SURF_ID='INERT'/ Furniture_Chimney
&OBST XB=14.70,15.20,11.40,13.00,0,2.40, SURF_ID='Brick'/ Furniture_FirePl
&OBST XB=14.60,14.70,9.80,11.40,0,1.00, SURF_ID='Brick'/ Furniture_FirePlWall
&OBST XB=9.66,10.28,11.57,14.68,0,1.00, RGB=153,102,0, SURF_ID='INERT'/ Furniture_KitchCounter
&OBST XB=10.28,14.63,14.07,14.68,0,1.00, RGB=153,102,0, SURF_ID='INERT'/ Furniture_KitchCounter
&OBST XB=13.70,14.23,2.58,3.18,0,1.00, RGB=153,102,0, SURF_ID='INERT'/ Furniture_Dresser
&OBST XB=10.70,11.23,2.58,3.18,0,1.00, RGB=153,102,0, SURF_ID='INERT'/ Furniture_Dresser
&OBST XB=16.60,16.70,3.00,3.10,0,0.80, RGB=153,102,0, SURF_ID='INERT'/ Furniture_TableDen
&OBST XB=16.60,18.20,3.00,3.90,0.80,0.90, RGB=153,102,0, SURF_ID='INERT'/ Furniture_TableDen
&OBST XB=16.60,16.70,3.80,3.90,0,0.80, RGB=153,102,0, SURF_ID='INERT'/ Furniture_TableDen
&OBST XB=18.10,18.20,3.00,3.10,0,0.80, RGB=153,102,0, SURF_ID='INERT'/ Furniture_TableDen
&OBST XB=18.10,18.20,3.80,3.90,0,0.80, RGB=153,102,0, SURF_ID='INERT'/ Furniture_TableDen
&OBST XB=26.56,27.99,0.79,1.19,2.00E-3,0.72, RGB=153,102,0, SURF_ID='INERT'/ Furniture_TableDen
&OBST XB=25.22,25.60,10.24,11.74,2.00E-3,0.72, RGB=153,102,0, SURF_ID='INERT'/ Furniture_TableDen
&OBST XB=28.10,30.05,2.00,3.30,0,0.3, RGB=51,51,255, SURF_ID='INERT'/ Furniture_Bed
&OBST XB=28.30,30.30,10.35,11.65,0,0.3, RGB=51,51,255, SURF_ID='INERT'/ Furniture_Bed
&OBST XB=29.60,30.20,9.50,10.10,0,0.50, RGB=153,102,0, SURF_ID='INERT'/ Furniture_EndTables
&OBST XB=15.36,15.89,6.70,7.30,0,0.50, RGB=153,102,0, SURF_ID='INERT'/ Furniture_EndTables
&OBST XB=18.93,19.47,6.70,7.30,0,0.50, RGB=153,102,0, SURF_ID='INERT'/ Furniture_EndTables
&OBST XB=18.93,19.47,6.70,7.30,0,0.50, RGB=153,102,0, SURF_ID='INERT'/ Furniture_EndTables
&OBST XB=29.60,30.20,11.84,12.44,0,0.50, RGB=153,102,0, SURF_ID='INERT'/ Furniture_EndTables
&OBST XB=16.47,18.47,7.20,7.30,0.50,0.90, COLOR='GRAY 60', SURF_ID='INERT'/ Furniture_Sofa
&OBST XB=16.47,16.60,6.70,7.20,0.50,0.60, COLOR='GRAY 60', SURF_ID='INERT'/ Furniture_Sofa
&OBST XB=18.33,18.47,6.70,7.20,0.50,0.60, COLOR='GRAY 60', SURF_ID='INERT'/ Furniture_Sofa
&OBST XB=16.47,18.47,6.70,7.30,0,0.50, COLOR='GRAY 60', SURF_ID='INERT'/ Furniture_Sofa
&OBST XB=30.10,30.23,14.76,16.76,0.50,0.90, COLOR='GRAY 60', SURF_ID='INERT'/ Furniture_Sofa
&OBST XB=29.57,30.10,16.66,16.76,0.50,0.60, COLOR='GRAY 60', SURF_ID='INERT'/ Furniture_Sofa
&OBST XB=29.57,30.10,14.76,14.86,0.50,0.60, COLOR='GRAY 60', SURF_ID='INERT'/ Furniture_Sofa
&OBST XB=29.57,30.23,14.76,16.76,0,0.50, COLOR='GRAY 60', SURF_ID='INERT'/ Furniture_Sofa
&OBST XB=16.00,16.10,3.00,3.40,0.50,0.60, RGB=153,153,255, SURF_ID='INERT'/ Furniture_Chair
&OBST XB=15.30,16.10,2.90,3.00,0.50,0.80, RGB=153,153,255, SURF_ID='INERT'/ Furniture_Chair
&OBST XB=15.30,15.40,3.00,3.40,0.50,0.60, RGB=153,153,255, SURF_ID='INERT'/ Furniture_Chair
&OBST XB=15.30,16.10,2.90,3.40,0,0.50, RGB=153,153,255, SURF_ID='INERT'/ Furniture_Chair
&OBST XB=19.50,19.60,3.00,3.40,0.50,0.60, RGB=153,153,255, SURF_ID='INERT'/ Furniture_Chair
&OBST XB=18.80,19.60,2.90,3.00,0.50,0.80, RGB=153,153,255, SURF_ID='INERT'/ Furniture_Chair
&OBST XB=18.80,18.90,3.00,3.40,0.50,0.60, RGB=153,153,255, SURF_ID='INERT'/ Furniture_Chair
&OBST XB=18.80,19.60,2.90,3.40,0,0.50, RGB=153,153,255, SURF_ID='INERT'/ Furniture_Chair
&OBST XB=11.30,13.00,4.00,6.50,0.70,0.80, RGB=153,102,0, SURF_ID='INERT'/ Furniture_DiningTable
&OBST XB=12.90,13.00,6.40,6.50,0,0.70, RGB=153,102,0, SURF_ID='INERT'/ Furniture_DiningTable
&OBST XB=11.30,11.40,6.40,6.50,0,0.70, RGB=153,102,0, SURF_ID='INERT'/ Furniture_DiningTable
&OBST XB=12.90,13.00,4.00,4.10,0,0.70, RGB=153,102,0, SURF_ID='INERT'/ Furniture_DiningTable
&OBST XB=11.30,11.40,4.00,4.10,0,0.70, RGB=153,102,0, SURF_ID='INERT'/ Furniture_DiningTable
&OBST XB=19.40,19.50,9.50,9.60,0,0.1000, SURF_ID='Roof Sheathing'/ Furniture_Couch1
&OBST XB=20.20,20.30,9.50,9.60,0,0.1000, SURF_ID='Roof Sheathing'/ Furniture_Couch1
&OBST XB=20.20,20.30,11.20,11.30,0,0.1000, SURF_ID='Roof Sheathing'/ Furniture_Couch1
&OBST XB=19.40,19.50,11.20,11.30,0,0.1000, SURF_ID='Roof Sheathing'/ Furniture_Couch1
&OBST XB=19.40,20.20,9.60,11.20,0.1000,0.50, SURF_ID='Sofa'/ Furniture_Couch1
&OBST XB=19.40,20.30,9.50,9.60,0.1000,0.70, SURF_ID='Sofa'/ Furniture_Couch1
&OBST XB=19.40,20.30,11.20,11.30,0.1000,0.70, SURF_ID='Sofa'/ Furniture_Couch1
&OBST XB=20.20,20.30,9.60,11.20,0.1000,0.90, SURF_ID='Sofa'/ Furniture_Couch1
&OBST XB=18.30,18.40,13.10,13.20,0,0.1000, SURF_ID='Sofa'/ Furniture_Couch2
&OBST XB=18.30,18.40,13.90,14.00,0,0.1000, SURF_ID='Sofa'/ Furniture_Couch2
&OBST XB=16.60,16.70,13.90,14.00,0,0.1000, SURF_ID='Sofa'/ Furniture_Couch2
&OBST XB=16.60,16.70,13.10,13.20,0,0.1000, SURF_ID='Sofa'/ Furniture_Couch2
&OBST XB=16.70,18.30,13.10,13.90,0.1000,0.50, SURF_ID='Sofa'/ Furniture_Couch2
&OBST XB=18.30,18.40,13.10,14.00,0.1000,0.70, SURF_ID='Sofa'/ Furniture_Couch2
&OBST XB=16.60,16.70,13.10,14.00,0.1000,0.70, SURF_ID='Sofa'/ Furniture_Couch2
&OBST XB=16.70,18.30,13.90,14.00,0.1000,0.90, SURF_ID='Sofa'/ Furniture_Couch2
&OBST XB=21.70,22.40,7.40,8.10,2.40,2.50, COLOR='INVISIBLE', PERMIT_HOLE=.FALSE., SURF_ID='Ceiling Gypsum', DEVC_ID='TIMER'/
E36Pull
&OBST XB=33.60,33.70,5.60,9.10,2.90,3.00, SURF_ID='Glass'/ Window_Steam
&OBST XB=33.60,33.70,5.80,8.90,3.00,3.10, SURF_ID='Glass'/ Window_Steam
&OBST XB=33.60,33.70,5.90,8.80,3.10,3.20, SURF_ID='Glass'/ Window_Steam
&OBST XB=33.60,33.70,6.10,8.60,3.20,3.30, SURF_ID='Glass'/ Window_Steam
&OBST XB=33.60,33.70,6.30,8.40,3.30,3.40, SURF_ID='Glass'/ Window_Steam
```

```
&OBST XB=33.60,33.70,6.40,8.30,3.40,3.50, SURF_ID='Glass'/ Window_Steam
&OBST XB=33.60,33.70,6.60,8.10,3.50,3.60, SURF_ID='Glass'/ Window_Steam
&OBST XB=33.60,33.70,6.70,8.00,3.60,3.70, SURF_ID='Glass'/ Window_Steam
&OBST XB=33.60,33.70,6.90,7.80,3.70,3.80, SURF_ID='Glass'/ Window_Steam
&OBST XB=33.60,33.70,7.00,7.70,3.80,3.90, SURF_ID='Glass'/ Window_Steam
&OBST XB=33.60,33.70,7.20,7.50,3.90,4.00, SURF_ID='Glass'/ Window_Steam
&OBST XB=33.60,33.70,5.40,9.30,2.80,2.90, SURF_ID='Glass'/ Window_Steam
&OBST XB=33.60,33.70,5.30,9.40,2.70,2.80, SURF_ID='Glass'/ Window_Steam
&OBST XB=11.30,12.40,14.90,15.00,1.20,2.10, PERMIT_HOLE=.FALSE., SURF_ID='Glass'/ Window_Kitchen Sink
&OBST XB=30.70,30.80,10.00,11.50,1.00,2.00, PERMIT_HOLE=.FALSE., SURF_ID='Glass'/ Window_Master
&OBST XB=16.10,18.90,2.10,2.20,1.00,2.20, PERMIT_HOLE=.FALSE., SURF_ID='Glass'/ Window_Living Room
&OBST XB=5.90,6.60,14.90,15.00,0.60,1.40, PERMIT_HOLE=.FALSE., SURF_ID='Glass'/ Window_west bathroom
&OBST XB=3.30,4.10,14.90,15.00,0.60,1.40, PERMIT_HOLE=.FALSE., SURF_ID='Glass'/ Window_Utility
&OBST XB=7.50,9.30,14.90,15.00,0,2.30, PERMIT_HOLE=.FALSE., SURF_ID='Glass'/ Window_Kitchen Slider
&OBST XB=23.20,23.90,17.40,17.50,0.60,1.40, PERMIT_HOLE=.FALSE., SURF_ID='Glass'/ Window_ExerciseSmall
&OBST XB=33.30,33.40,5.30,6.20,0.80,1.60, PERMIT_HOLE=.FALSE., SURF_ID='Glass'/ Window_Sauna
&OBST XB=27.50,29.30,0.40,0.50,1.30,2.00, PERMIT_HOLE=.FALSE., SURF_ID='Glass'/ Window_Front_Bedroom
&OBST XB=22.10,22.70,3.40,3.50,0.40,2.00, PERMIT_HOLE=.FALSE., SURF_ID='Glass'/ Window_Front Door Left
&OBST XB=30.70,30.80,3.00,3.70,0.60,1.40, RGB=0,204,204, TRANSPARENCY=0.40, PERMIT_HOLE=.FALSE., SURF_ID='Glass'/
Window_Greenhouse
&OBST XB=25.80,29.40,17.40,17.50,0,2.30, PERMIT_HOLE=.FALSE., SURF_ID='Glass'/ Window_ExerciseBig
&OBST XB=0.2000,0.3,8.00,9.80,0.60,1.40, PERMIT_HOLE=.FALSE., SURF_ID='Glass'/ Window_SideGarage
&OBST XB=11.10,13.00,2.10,2.20,1.00,2.20, PERMIT_HOLE=.FALSE., SURF_ID='Glass'/ Window_Dining Room
&OBST XB=24.10,24.70,3.40,3.50,0.40,2.00, PERMIT_HOLE=.FALSE., SURF_ID='Glass'/ Window_Front Door Right
&OBST XB=15.30,16.20,13.60,14.90,2.30,2.40, PERMIT_HOLE=.FALSE., SURF_ID='Glass', DEVC_ID='TIMER22'/ Solarium_A_Panels
&OBST XB=15.30,16.20,14.90,15.00,-0.1,2.40, PERMIT_HOLE=.FALSE., SURF_ID='Glass', DEVC_ID='TIMER22'/ Solarium_A_Panels
&OBST XB=19.20,20.20,14.90,15.00,-0.1,2.40, PERMIT_HOLE=.FALSE., SURF_ID='Glass', DEVC_ID='TIMER3'/ Solarium_E_Panels
&OBST XB=19.20,20.20,13.60,14.90,2.30,2.40, PERMIT_HOLE=.FALSE., SURF_ID='Glass', DEVC_ID='TIMER4'/ Solarium_1_Panels
&OBST XB=18.20,19.20,14.90,15.00,-0.1,2.40, PERMIT_HOLE=.FALSE., SURF_ID='Glass', DEVC_ID='TIMER3'/ Solarium_D_Panels
&OBST XB=18.20,19.20,13.60,14.90,2.30,2.40, PERMIT_HOLE=.FALSE., SURF_ID='Glass', DEVC_ID='TIMER3'/ Solarium_D_Panels
&OBST XB=17.20,18.20,14.90,15.00,-0.1,2.40, PERMIT_HOLE=.FALSE., SURF_ID='Glass', DEVC_ID='TIMER3'/ Solarium_C_Panels
&OBST XB=17.20,18.20,13.60,14.90,2.30,2.40, PERMIT_HOLE=.FALSE., SURF_ID='Glass', DEVC_ID='TIMER3'/ Solarium_C_Panels
&OBST XB=16.20,17.20,14.90,15.00,-0.1,2.40, PERMIT_HOLE=.FALSE., SURF_ID='Glass', DEVC_ID='TIMER22'/ Solarium_B_Panels
&OBST XB=16.20,17.20,13.60,14.90,2.30,2.40, PERMIT_HOLE=.FALSE., SURF_ID='Glass', DEVC_ID='TIMER22'/ Solarium_B_Panels
&OBST XB=15.30,20.20,13.50,13.60,2.40,2.50, SURF_ID='Glass'/ SolariumGlassTop
&OBST XB=2.70,14.70,10.00,15.00,-0.1,0, SURF_ID='Wood Floor'/ Flooring_Kitchen
&OBST XB=9.20,14.70,7.90,10.00,-0.1,0, SURF_ID='Wood Floor'/ Flooring_Kitchen
&OBST XB=12.60,14.70,7.60,7.90,-0.1,0, SURF_ID='Wood Floor'/ Flooring_Pantry
&OBST XB=14.70,21.30,7.60,13.70,-0.1,0, SURF_ID='Wood Floor'/ Flooring_Den
&OBST XB=21.30,21.90,9.00,13.70,-0.1,0, SURF_ID='Wood Floor'/ Flooring_Den
&OBST XB=14.70,20.20,13.70,15.00,-0.1,0, SURF_ID='Tile'/ Flooring_Solarium
&OBST XB=10.10,24.90,2.10,7.60,-0.1,0, SURF_ID='Carpet'/ Flooring_LivDiningRm
&OBST XB=10.10,12.60,7.60,7.90,-0.1,0, SURF_ID='Carpet'/ Flooring_DiningRm
&OBST XB=21.30,24.90,7.60,9.00,-0.1,0, SURF_ID='Carpet'/ Flooring_Hall
&OBST XB=20.20,30.80,13.70,17.50,-0.1,0, SURF_ID='Carpet'/ Flooring_ExerciseRm
&OBST XB=24.90,30.70,7.80,13.70,-0.1,0, SURF_ID='Carpet'/ Flooring_MasterBed
&OBST XB=24.90,30.70,6.30,7.80,-0.1,0, SURF_ID='Carpet'/ Flooring_BathBed
&OBST XB=24.90,27.90,0.40,6.30,-0.1,0, SURF_ID='Carpet'/ Flooring_FrontBedroom
&OBST XB=27.90,30.70,0.40,4.70,-0.1,0, SURF_ID='Carpet'/ Flooring_FrontBedroom
&OBST XB=30.70,33.40,7.60,9.90,-0.1,0, SURF_ID='Carpet'/ Flooring_MBedCloset
&OBST XB=27.90,30.70,4.70,6.30,-0.1,0, SURF_ID='Tile'/ Flooring_FrontBath
&OBST XB=30.70,33.40,4.70,7.60,-0.1,0, SURF_ID='Tile'/ Flooring_SteamRm
&OBST XB=21.90,24.90,9.00,13.70,-0.1,0, SURF_ID='Tile'/ Flooring_MasterBath
&OBST XB=2.70,10.10,2.10,8.00,-0.1,0, SURF_ID='Concrete'/ Flooring_Garage
&OBST XB=2.70,9.20,8.00,10.00,-0.1,0, SURF_ID='Concrete'/ Flooring_Garage
&OBST XB=0.2000,2.70,2.10,12.10,-0.1,0, SURF_ID='Concrete'/ Flooring_Workshop
&OBST XB=30.70,33.40,0.40,4.70,-0.1,0, SURF_ID='Concrete'/ Flooring_GreenHouse
&OBST XB=2.70,2.80,5.40,6.30,0.1000,2.00, PERMIT_HOLE=.FALSE., SURF_ID='Wood Joists'/ Door_Front Garage Storage
&OBST XB=25.70,27.30,4.70,4.80,0.1000,2.00, COLOR='GRAY 80', PERMIT_HOLE=.FALSE., SURF_ID='INTERIORDOOR'/ Doors_ ABedCloset
&OBST XB=24.20,25.10,4.70,4.80,0.1000,2.00, COLOR='GRAY 80', PERMIT_HOLE=.FALSE., SURF_ID='INTERIORDOOR'/ Doors_ ABed
&OBST XB=23.00,23.90,3.40,3.50,0.1000,2.00, COLOR='GRAY 80', PERMIT_HOLE=.FALSE., SURF_ID='Roof Sheathing', DEVC_ID='TIMER5'/
Doors_ Front Door
&OBST XB=2.70,2.80,12.80,13.60,0.1000,2.00, COLOR='GRAY 80', PERMIT_HOLE=.FALSE., SURF_ID='Roof Sheathing'/ Doors_ PoolsideExit
&OBST XB=1.20,2.10,2.10,2.20,0.1000,2.00, COLOR='GRAY 80', PERMIT_HOLE=.FALSE., SURF_ID='Roof Sheathing'/ Doors_ WorkroomExit
&OBST XB=14.80,15.10,12.00,12.40,0.50,0.60, PERMIT_HOLE=.FALSE., SURF_ID='INERT'/ Obstruction
&HOLE XB=2.96E1,3.00E1,7.90,8.39,3.40,3.81, COLOR='INVISIBLE'/ RoofVent6
&HOLE XB=3.36E1,3.37E1,7.30,7.40,2.70,4.00, COLOR='INVISIBLE'/ Wall_Interior
&HOLE XB=1.18E1,1.28E1,8.90,9.90,2.40,3.60, COLOR='INVISIBLE'/ Hole
&HOLE XB=1.18E1,1.29E1,9.90,1.00E1,3.40,3.50, COLOR='INVISIBLE'/ Hole
&HOLE XB=1.28E1,1.29E1,9.30,9.90,3.40,3.50, COLOR='INVISIBLE'/ Hole
&HOLE XB=9.80,1.02E1,1.09E1,1.14E1,3.50,3.90, COLOR='INVISIBLE'/ Hole
&HOLE XB=2.15E1,2.49E1,2.10,3.40,-1.00E-1,0, COLOR='INVISIBLE'/ Hole_FrontStep
&HOLE XB=2.17E1,2.24E1,7.40,8.10,2.20,2.65, COLOR='INVISIBLE'/ Hole_E36 Ceiling Pull
&HOLE XB=2.845E1,2.845E1,5.90,6.10,2.39,2.51, COLOR='INVISIBLE'/ HoleRegister_DBath
&HOLE XB=2.28E1,2.30E1,5.40,5.50,2.39,2.51, COLOR='INVISIBLE'/ HoleRegister_FrontDoor
&HOLE XB=1.59E1,1.60E1,7.60,7.70,3.00,3.10, COLOR='INVISIBLE'/ HoleRegister_Den
&HOLE XB=1.60E1,1.61E1,7.60,7.70,3.00,3.10, COLOR='INVISIBLE'/ HoleRegister_Den
&HOLE XB=1.61E1,1.62E1,7.60,7.70,3.00,3.10, COLOR='INVISIBLE'/ HoleRegister_Den
&HOLE XB=1.61E1,1.62E1,7.60,7.70,3.10,3.20, COLOR='INVISIBLE'/ HoleRegister_Den
&HOLE XB=1.60E1,1.61E1,7.60,7.70,3.10,3.20, COLOR='INVISIBLE'/ HoleRegister_Den
&HOLE XB=1.59E1,1.60E1,7.60,7.70,3.10,3.20, COLOR='INVISIBLE'/ HoleRegister_Den
&HOLE XB=2.16E1,2.18E1,9.70,1.01E1,2.39,2.51, COLOR='INVISIBLE'/ HoleRegister_Den
&HOLE XB=1.48E1,1.51E1,1.20E1,1.24E1,0,4.00, COLOR='INVISIBLE'/ Hole_Chimney
&HOLE XB=1.51E1,1.52E1,1.17E1,1.27E1,0,5.00E-1, COLOR='INVISIBLE'/ Hole_Chimney
&HOLE XB=1.01E1,1.02E1,6.80,7.50,1.00E-1,3.00E-1, COLOR='INVISIBLE'/ Hole_FurnaceReturn
&HOLE XB=2.73E1,2.75E1,4.60,4.70,2.39,2.51, COLOR='INVISIBLE'/ HoleRegister_GuestBed
&HOLE XB=2.73E1,2.75E1,5.00E-1,6.00E-1,2.39,2.51, COLOR='INVISIBLE'/ HoleRegister_GuestBed
&HOLE XB=2.56E1,2.58E1,7.80,7.90,2.39,2.51, COLOR='INVISIBLE'/ HoleRegister_MasterBed
&HOLE XB=2.56E1,2.58E1,1.37E1,1.38E1,2.39,2.51, COLOR='INVISIBLE'/ HoleRegister_Exercise
&HOLE XB=2.20E1,2.22E1,1.37E1,1.38E1,2.39,2.51, COLOR='INVISIBLE'/ HoleRegister_Exercise
&HOLE XB=1.59E1,1.62E1,7.50,7.60,2.39,2.51, COLOR='INVISIBLE'/ HoleRegister_LivRoom
&HOLE XB=1.85E1,1.88E1,7.50,7.60,2.39,2.51, COLOR='INVISIBLE'/ HoleRegister_LivRoom
&HOLE XB=1.40E1,1.42E1,7.50,7.60,2.39,2.51, COLOR='INVISIBLE'/ HoleRegister_DinRoom
&HOLE XB=1.45E1,1.46E1,1.18E1,1.20E1,2.39,2.51, COLOR='INVISIBLE'/ HoleRegister_Kitch
&HOLE XB=1.06E1,1.08E1,1.19E1,1.20E1,2.39,2.51, COLOR='INVISIBLE'/ HoleRegister_Kitch
```

```
&HOLE XB=5.70,5.90,1.19E1,1.20E1,2.39,2.51, COLOR='INVISIBLE'/ HoleRegister_Pantry
&HOLE XB=1.90,2.10,1.155E1,1.165E1,2.39,2.51, COLOR='INVISIBLE'/ HoleRegister_Workshop
&HOLE XB=1.89E1,2.08E1,1.50,1.60,2.30,2.60, COLOR='INVISIBLE'/ HoleSoffit Vent A1
&HOLE XB=-4.00E-1,-3.00E-1,5.30,7.20,2.3,2.60, COLOR='INVISIBLE'/ HoleSoffit Vent B1
&HOLE XB=2.70,4.60,1.50,1.60,2.30,2.60, COLOR='INVISIBLE'/ HoleSoffit Vent A4
&HOLE XB=2.745E1,2.935E1,1.79E1,1.80E1,2.30,2.60, COLOR='INVISIBLE'/ HoleSoffit Vent C4
&HOLE XB=1.35E1,1.54E1,1.50,1.60,2.30,2.60, COLOR='INVISIBLE'/ HoleSoffit Vent A2
&HOLE XB=2.205E1,2.395E1,1.79E1,1.80E1,2.30,2.60, COLOR='INVISIBLE'/ HoleSoffit Vent C3
&HOLE XB=8.10,1.00E1,1.50,1.60,2.30,2.60, COLOR='INVISIBLE'/ HoleSoffit Vent A3
&HOLE XB=1.055E1,1.245E1,1.58E1,1.59E1,2.30,2.60, COLOR='INVISIBLE'/ HoleSoffit Vent C2
&HOLE XB=5.15,7.05,1.58E1,1.59E1,2.30,2.60, COLOR='INVISIBLE'/ HoleSoffit Vent C1
&HOLE XB=2.02E1,2.05E1,4.40,5.20,2.20,2.70, COLOR='INVISIBLE'/ HoleFFPunchDown
&HOLE XB=2.08E1,2.11E1,4.40,5.20,2.20,2.70, COLOR='INVISIBLE'/ HoleFFPunchDown2
&HOLE XB=1.53E1,2.02E1,1.47E1,1.52E1,0,2.30, COLOR='INVISIBLE'/ HoleSolarium
&HOLE XB=2.41E1,2.47E1,3.30,3.60,4.00E-1,2.00, COLOR='INVISIBLE'/ HoleFront Door Right Window
&HOLE XB=3.06E1,3.10E1,1.00E1,1.15E1,1.00,2.00, COLOR='INVISIBLE'/ HoleMaster Window
&HOLE XB=1.61E1,1.89E1,1.90,2.40,1.00,2.20, COLOR='INVISIBLE'/ HoleLiving Room Window
&HOLE XB=5.90,6.60,1.47E1,1.51E1,6.00E-1,1.40, COLOR='INVISIBLE'/ HoleWest bathroom window
&HOLE XB=3.30,4.10,1.47E1,1.52E1,6.00E-1,1.40, COLOR='INVISIBLE'/ HoleUtility Window
&HOLE XB=2.32E1,2.39E1,1.73E1,1.76E1,6.00E-1,1.40, COLOR='INVISIBLE'/ HoleExercise Window
&HOLE XB=3.32E1,3.36E1,5.30,6.20,8.00E-1,1.60, COLOR='INVISIBLE'/ HoleSauna Window
&HOLE XB=3.06E1,3.09E1,3.00,3.70,6.00E-1,1.40, COLOR='INVISIBLE'/ HoleFront bedroom Greenhouse
&HOLE XB=2.75E1,2.93E1,3.00E-1,6.00E-1,1.30,2.00, COLOR='INVISIBLE'/ HoleFront_Bedroom Window
&HOLE XB=2.21E1,2.27E1,3.30,3.60,4.00E-1,2.00, COLOR='INVISIBLE'/ HoleFront Door Left Window
&HOLE XB=2.58E1,2.94E1,1.73E1,1.76E1,0,2.30, COLOR='INVISIBLE'/ HoleExercise Window Wall
&HOLE XB=7.50,9.30,1.47E1,1.52E1,0,2.30, COLOR='INVISIBLE'/ HoleGlass slider kitchen
&HOLE XB=1.13E1,1.24E1,1.47E1,1.52E1,1.20,2.10, COLOR='INVISIBLE'/ HoleKitchen Sink Window
&HOLE XB=1.11E1,1.30E1,1.90,2.40,1.00,2.20, COLOR='INVISIBLE'/ HoleDinning Room Window
&HOLE XB=0.0,5.00E-1,8.00,9.80,6.00E-1,1.40, COLOR='INVISIBLE'/ HoleSide Garage Window
&HOLE XB=2.57E1,2.73E1,4.60,4.90,0,2.00, COLOR='INVISIBLE'/ HoleCloset Door Front Bedroom
&HOLE XB=2.60,2.90,1.28E1,1.36E1,0,2.00, COLOR='INVISIBLE'/ HoleB Side Exterior Doorway
&HOLE XB=3.06E1,3.10E1,7.80,8.70,0,2.00, COLOR='INVISIBLE'/ HoleMB to Bath Doorway
&HOLE XB=3.90,8.90,2.10,2.40,0,2.00, COLOR='INVISIBLE'/ HoleGarage Door
&HOLE XB=2.42E1,2.51E1,4.60,4.90,0,2.00, COLOR='INVISIBLE'/ HoleFront_Right Bedroom Door
&HOLE XB=2.13E1,2.16E1,4.00,5.10,0,2.00, COLOR='INVISIBLE'/ HoleOpening from Fdoor to Living Room
&HOLE XB=1.45E1,1.48E1,4.20,6.30,0,2.00, COLOR='INVISIBLE'/ Holeopening living to dining
&HOLE XB=1.32E1,1.41E1,8.70,9.00,0,2.00, COLOR='INVISIBLE'/ HolePantry door
&HOLE XB=3.06E1,3.09E1,5.30,6.20,0,2.00, COLOR='INVISIBLE'/ HoleRight bathroom
&HOLE XB=2.30E1,2.39E1,3.30,3.60,0,2.00, COLOR='INVISIBLE'/ HoleFront Door
&HOLE XB=2.71E1,2.89E1,1.35E1,1.38E1,0,2.00, COLOR='INVISIBLE'/ HoleMasterBed to Exercise
&HOLE XB=2.31E1,2.40E1,1.34E1,1.39E1,0,2.00, COLOR='INVISIBLE'/ HoleCenter bathroom Rear Door
&HOLE XB=3.06E1,3.09E1,6.50,7.40,0,2.00, COLOR='INVISIBLE'/ Holeright bathroom door2
&HOLE XB=2.31E1,2.40E1,8.80,9.10,0,2.00, COLOR='INVISIBLE'/ HoleBathroom Door
&HOLE XB=2.58E1,2.67E1,7.80,7.80,0,2.00, COLOR='INVISIBLE'/ HoleMasterbedroom Closet
&HOLE XB=2.34E1,2.43E1,7.60,7.80,0,2.00, COLOR='INVISIBLE'/ HoleFire Room Door
&HOLE XB=2.90E1,2.99E1,4.50,5.00,0,2.00, COLOR='INVISIBLE'/ HoleFront Bedroom bathroom
&HOLE XB=5.70,6.60,9.70,1.02E1,0,2.00, COLOR='INVISIBLE'/ HoleInto Garage
&HOLE XB=2.50,3.00,8.00,8.90,0,2.00, COLOR='INVISIBLE'/ HoleRear Garage Storage
&HOLE XB=1.20,2.10,1.90,2.40,0,2.00, COLOR='INVISIBLE'/ HoleDoor_Garage Acces Outside
&HOLE XB=5.70,6.60,1.25E1,1.30E1,0,2.00, COLOR='INVISIBLE'/ HoleRear Bathroom
&HOLE XB=2.50,3.00,5.40,6.30,0,2.00, COLOR='INVISIBLE'/ HoleFront Garage Storage
&HOLE XB=2.30E1,2.37E1,9.00,1.02E1,2.30,3.80, COLOR='INVISIBLE'/ HoleSkylight
&HOLE XB=1.99E1,2.13E1,4.20,5.40,3.20,4.30, COLOR='INVISIBLE'/ HoleRoof Cut
&HOLE XB=2.37E1,2.42E1,6.50,7.00,2.30,2.80, COLOR='INVISIBLE'/ BurnerHole
&VENT SURF_ID='CouchBurner', XB=19.39,19.39,9.60,11.20,0.1000,0.50/ Furniture_Couch1
&VENT SURF_ID='CouchBurner', XB=19.40,20.20,9.60,11.20,0.51,0.51/ Furniture_Couch1
&VENT SURF_ID='CouchBurner', XB=20.19,20.19,9.60,11.20,0.50,0.90/ Furniture_Couch1
&VENT SURF_ID='CouchBurner', XB=20.31,20.31,9.60,11.20,0.1000,0.50/ Furniture_Couch1
&VENT SURF_ID='CouchBurner', XB=19.40,20.30,9.49,9.49,0.1000,0.70/ Furniture_Couch1
&VENT SURF_ID='CouchBurner', XB=19.40,20.30,11.31,11.31,0.1000,0.70/ Furniture_Couch1
&VENT SURF_ID='CouchBurner2', XB=16.70,18.30,13.09,13.09,0.1000,0.50/ Furniture_Couch2
&VENT SURF_ID='CouchBurner2', XB=16.70,18.30,13.10,13.90,0.51,0.51/ Furniture_Couch2
&VENT SURF_ID='CouchBurner2', XB=16.70,18.30,13.89,13.89,0.50,0.90/ Furniture_Couch2
&VENT SURF_ID='CouchBurner2', XB=16.70,18.30,14.01,14.01,0.1000,0.90/ Furniture_Couch2
&VENT SURF_ID='CouchBurner2', XB=18.41,18.41,13.10,14.00,0.1000,0.70/ Furniture_Couch2
&VENT SURF_ID='CouchBurner2', XB=16.59,16.59,13.10,14.00,0.1000,0.70/ Furniture_Couch2
&VENT SURF_ID='OPEN', XB=-21.40,53.00,-10.00,-10.00,-0.1,11.00, COLOR='INVISIBLE'/ BndVent
&VENT SURF_ID='INERT', XB=-21.40,53.00,-10.00,27.20,-0.1,-0.1, COLOR='INVISIBLE'/ BndVent
&VENT SURF_ID='OPEN', XB=-21.40,53.00,-10.00,27.20,11.00,11.00, COLOR='INVISIBLE'/ BndVent
&VENT SURF_ID='OPEN', XB=-21.40,-21.40,-10.00,27.20,-0.1,11.00, COLOR='INVISIBLE'/ BndVent
&VENT SURF_ID='OPEN', XB=53.00,53.00,-10.00,27.20,-0.1,11.00, COLOR='INVISIBLE'/ BndVent
&VENT SURF_ID='Wind', XB=-21.40,53.00,27.20,27.20,-0.1,11.00, IOR=-2, COLOR='INVISIBLE'/ BndVent
&BNDF QUANTITY='MASS FLUX', SPEC_ID='fuel'/
&ISOF QUANTITY='VOLUME FRACTION', SPEC_ID='oxygen', VALUE=0.1500/
&SLCF QUANTITY='TEMPERATURE', PBX=18.00/
&SLCF QUANTITY='VELOCITY', VECTOR=.TRUE., PBX=18.00/
&SLCF QUANTITY='VOLUME FRACTION', SPEC_ID='fuel', PBX=19.50/
&SLCF QUANTITY='VELOCITY', VECTOR=.TRUE., PBX=19.80/
&SLCF QUANTITY='VOLUME FRACTION', SPEC_ID='oxygen', PBX=19.90/
&SLCF QUANTITY='VOLUME FRACTION', SPEC_ID='oxygen', PBX=22.20/
&SLCF QUANTITY='TEMPERATURE', VECTOR=.TRUE., PBX=22.20/
&SLCF QUANTITY='TEMPERATURE', VECTOR=.TRUE., PBX=23.50/
&SLCF QUANTITY='VOLUME FRACTION', SPEC_ID='oxygen', PBY=10.50/
&SLCF QUANTITY='TEMPERATURE', VECTOR=.TRUE., PBY=4.60/
&SLCF QUANTITY='TEMPERATURE', PBY=8.30/
&SLCF QUANTITY='VOLUME FRACTION', SPEC_ID='oxygen', PBY=8.50/
&SLCF QUANTITY='PRESSURE', PBZ=1.00/
&SLCF QUANTITY='TEMPERATURE', PBZ=1.00/
&SLCF QUANTITY='TEMPERATURE', VECTOR=.TRUE., PBZ=1.50/
&SLCF QUANTITY='VELOCITY', VECTOR=.TRUE., PBZ=1.50/
&SLCF QUANTITY='VELOCITY', VECTOR=.TRUE., PBZ=2.20/
&SLCF QUANTITY='PRESSURE', PBZ=2.60/
&SLCF QUANTITY='VELOCITY', VECTOR=.TRUE., PBZ=2.80/
&TAIL /
```